WHITEHEAD'S

PHILOSOPHY

OF SCIENCE

WHITEHEAD'S
PHILOSOPHY
OF SCIENCE

By ROBERT M. PALTER

THE UNIVERSITY OF CHICAGO PRESS
Chicago and London

Standard Book Number: 226-64523-1
Library of Congress Catalog Card Number: 60-7241

THE UNIVERSITY OF CHICAGO PRESS, CHICAGO 60637
The University of Chicago Press, Ltd., London

Published 1960. Second Impression 1970

Printed in the United States of America

TO TONI

PREFACE

It is generally agreed that Whitehead's philosophy is extremely difficult; that his philosophy is also profound, or even important, is today a rare view indeed. Convinced as I am of both the difficulty and the profundity of Whitehead's philosophy, I have written this book in the hope of dispelling some of the apparent obscurities and exposing some of the real achievements in at least one area of Whitehead's philosophic activity, namely, his contributions to the philosophy of science. My method has been to give a concise but connected account of Whitehead's philosophy of science and to touch upon wider, "metaphysical" issues only where they directly impinge upon the analysis of science. It is scarcely necessary to say that my book is in no sense a substitute for any or all of Whitehead's writings on philosophy of science; on the contrary, much of what I say will be fully intelligible only to one who already possesses a first-hand acquaintance with Whitehead's philosophy. On the other hand, I trust that my exposition will stimulate some readers to a closer examination of Whitehead's philosophical works and at the same time provide a general framework (certainly neither foolproof nor unique) capable of organizing the deep insights into science and philosophy which, I am confident, will emerge from such an examination.

I wish to express my gratitude to Adolf Grünbaum, who read the original draft of this book with great care and insight and suggested many corrections and improvements which have been incorporated into the text; to Vere Chappell and Arthur Woodruff, who read and commented on parts of the manuscript and assisted in the final task of proofreading; to Jack McClurg for his considered judgment on a part of the manuscript. My greatest debt is to Howard Stein, who patiently read successive drafts of the entire book, correcting errors and making invaluable suggestions for changes in style and matter. I profited immensely from discussions with Mr. Stein of many of the critical issues in mathematics, physics, and philosophy which arose during my work on the book; his profound knowledge in each of these fields has helped me to understand many problematic passages in Whitehead's writings.

I am grateful to the College of the University of Chicago for a grant which helped defray the cost of preparing the manuscript.

I wish to thank the following publishers for permission to quote from the works of Whitehead and other authors: Cambridge University Press, Encyclopaedia Britannica, Alfred A. Knopf, the Library of Living Philosophers, the Macmillan Company, Philosophical Library, Princeton University Press, and Williams and Norgate. I am also indebted to the Editor of the Aristotelian Society, the Physical Society of London, the Rice Institute, and the Royal Society of London for permission to quote from their publications.

NOTE TO SECOND IMPRESSION

I should like to call attention to two oversights:

1. Page 108. There is a contradiction in Whitehead's assumptions about *connection* and *inclusion*. If A includes B, then A is connected with B and also every region connected with B is connected with A; so A is connected with itself, which contradicts one of the assumptions about connection.

2. Page 148, n. 4. Whitehead *does* call "simple location" a fallacy—not, to be sure, in his original discussion of the subject in SMW, but later in PR (p. 208).

My presentation of Whitehead's views on space and relativity has been subjected to a searching criticism by Adolf Grünbaum in a review essay, *Philosophical Review,* 71 (April, 1962): 218-29.

Whitehead's theory of relativity has been further studied by physicists since I compiled Appendix IV; I mention here only the most detailed comparison to date of Whitehead's theory with alternative theories of gravitation: G. J. Whitrow and G. E. Morduch, "Relativistic Theories of Gravitation," *Vistas in Astronomy,* vol. 6, ed. A. Beer (Oxford: Pergamon Press, 1965). I have myself given a simple, intuitive account of Whitehead's theory in "Science and Its History in the Philosophy of Whitehead," *The Hartshorne Festschrift: Process and Divinity,* ed. W. Reese and E. Freeman (LaSalle, Ill.: Open Court, 1964).

A summary of Whitehead's views on mathematics, deliberately ignored in the text, may be found in my "The Place of Mathematics in Whitehead's Philosophy," *Alfred North Whitehead: Essays on His Philosophy,* ed. G. Kline (Englewood Cliffs, N. J.: Prentice-Hall, Inc., 1963).

For a discussion of the relation between quantum physics and Whitehead's philosophy of organism, see A. Shimony, "Quantum Physics and the Philosophy of Whitehead," *Philosophy in America,* ed. M. Black (Ithaca, N. Y.: Cornell University Press, 1965). Also of interest in this connection is J. M. Burgers, *Experience and Conceptual Activity* (Cambridge: MIT Press, 1965).

CONTENTS

LIST OF FIGURES

NOTE ON ABBREVIATIONS

In citing Whitehead's writings I have used the following abbreviations throughout. Full publication data for all works cited may be found in the Bibliography.

ADG   *The Axioms of Descriptive Geometry*
AG   "The Axioms of Geometry" (in *Essays in Science and Philosophy*)
AI   *Adventures of Ideas*
APG   *The Axioms of Projective Geometry*
ASI   "The Anatomy of Some Scientific Ideas" (chap. ix in *The Aims of Education*)
CN   *The Concept of Nature*
ET   "Einstein's Theory" (in *Essays in Science and Philosophy*)
LPR   "Letter to the Editor" on *The Principle of Relativity*
M   "Analysis of Meaning" (in *Essays in Science and Philosophy*)
MC   "On Mathematical Concepts of the Material World"
MT   *Modes of Thought*
NG   (with B. Russell) "Non-Euclidean Geometry" (in *Essays in Science and Philosophy*)
PA   "The Philosophical Aspects of the Theory of Relativity"
PNK   *An Enquiry Concerning the Principles of Natural Knowledge*
PR   *Process and Reality*
PS   "The Problem of Simultaneity"
R   *The Principle of Relativity*
S   *Symbolism*
SMW   *Science and the Modern World*
TSM   "Time, Space, and Material: Are They, and If So in What Sense, the Ultimate Data of Science?"
UA   *Universal Algebra*
UC   "Uniformity and Contingency" (in *Essays in Science and Philosophy*)

When referring in one chapter to an equation from another chapter, I have used first the chapter number and then the equation number, e.g., (2.3) for equation (3) in chap. ii.

INTRODUCTION

Whitehead's philosophy of science has been viewed, at one extreme, as his sole important contribution to philosophy, an early achievement whose bright promise was unfortunately overshadowed by a series of increasingly unintelligible later works written under some not always identifiable but obviously pernicious influences; or, at the other extreme, as a series of essays in mathematical logic of no particular philosophical interest except insofar as they occasionally foreshadow the metaphysical categories and doctrines of the later works. Both of these extreme views agree in making Whitehead's "scientific" and "metaphysical" writings more or less independent of one another; they differ as to which strain in Whitehead's writings—the "scientific" or the "metaphysical"—is philosophically more important. For Whitehead himself the problem of the relation between philosophy of science and metaphysics is mainly a question of timing and of emphasis: while doing philosophy of science it is possible to ignore certain large metaphysical issues; and, though the converse is certainly not true (science being a vital datum for metaphysics), the results of an independent and prior analysis of science may be incorporated without essential change in the developed metaphysical system. On the other hand, certain types of philosophical considerations, traditionally termed "metaphysical," do bulk large in Whitehead's philosophy of science; moreover, special adjustments in the philosophy of science are required when it is incorporated within the metaphysical system of Whitehead's "philosophy of organism." We have, I would suggest, in Whitehead's philosophy an instructive example of the ways in which science and metaphysics may fruitfully interact with one another.

In his three early works on philosophy of science (PNK, 1919; CN, 1920; R, 1922), Whitehead insists upon the exclusion of all considerations of the type which he terms "metaphysical." The subject matter of philosophy of science is nature, construed as that complex of entities and relations which is revealed in sense-perception. According to Whitehead, nature is "self-contained for thought," i.e., the study of nature can be pursued without reference to the fact that nature is being thought about. Thus at the outset all concern with aesthetic

1

and moral values (which Whitehead holds to be indubitably present in nature)[1] is rejected; more generally, the ultimate status of the entities of natural science among all possible entities is to be entirely ignored. On the other hand, it is clear that the analysis of science has relevance for these larger issues; and, in retrospect, we can see Whitehead deliberately narrowing his horizon at first in preparation for the all-inclusive synthesis which is to follow.[2] He says:

We leave to metaphysics the synthesis of the knower and the known. . . . any metaphysical interpretation is an illegitimate importation into the philosophy of natural science. By a metaphysical interpretation I mean any discussion of the how (beyond nature) and of the why (beyond nature) of thought and sense-awareness [CN, p. 47].

But if metaphysics is irrelevant for philosophy of science, the converse, as noted above, is not the case:

We are not called upon to make any pronouncement as to the psychological relation of subjects to objects or as to the status of either in the realm of reality. It is true that the issue of our endeavour may provide material which is relevant evidence for a discussion of that question. It can hardly fail to do so. But it is only evidence, and is not itself the metaphysical discussion [CN, p. 47].

In his later works Whitehead gives us this metaphysical discussion at great length and with a profusion of detail which has tended to bewilder hostile and friendly critics alike. "Speculative philosophy," says Whitehead, "is the endeavour to frame a coherent, logical, necessary system of general ideas in terms of which every element of our experience can be interpreted" (PR, p. 4). Natural science, since it is an element of our experience, must certainly find a place— and indeed, as it turns out, a central place—in the system. An interesting question is, what happens to Whitehead's older philosophy of science in the new context? In general, there is considerable modification, much of which can be traced to the increased emphasis on percipient events and knowing minds. An example may make this clearer. In his earlier work Whitehead lays great stress upon an account of perception whose most unusual and important feature is that it asserts the possibility of an awareness of distant events in their character of bare relata with respect to the qualitatively perceived events nearby. Such a minimal awareness of spatially distant events is taken as a primitive, given fact in the early doctrine of perception whereas it is replaced in the philosophy of organism by an elaborate explanation of just how a percipient event manages

[1] In this connection Whitehead speaks of "excluding any reference to moral or aesthetic values whose apprehension is vivid in proportion to self-conscious activity. The values of nature are perhaps the key to the metaphysical synthesis of existence. But such a synthesis is exactly what I am not attempting" (CN, p. 5). Here "nature" includes values; sometimes in Whitehead's early writings "nature" excludes values, e.g.: "Nature is the system of factors apprehended in sense-awareness. . . . Divest consciousness of its ideality, such as its logical, emotional, aesthetic and moral apprehensions, and what is left is sense-awareness" (R, p. 20).

[2] Thus, as early as 1924, in the Preface to the second edition of PNK, we find Whitehead saying: "I hope in the immediate future to embody the standpoint of these volumes [namely, PNK, CN, and R] in a more complete metaphysical study" (p. ix).

to perceive its contemporary world in what is called the perceptual mode of "presentational immediacy." This later view, however, involves not merely a shift from philosophy of science to metaphysics but also a fundamental technical revision of the earlier version of the method of extensive abstraction—a revision which, according to Whitehead, enables him to define straight lines independently of measurement and thereby to explain presentational immediacy. Thus there is a subtle readjustment of ideas induced by the interaction between the diverse aspects of Whitehead's speculative philosophy.

A further illustration of the ways in which science and metaphysics are related in Whitehead's work is afforded by comparing his definition of speculative philosophy (already quoted) with his theory of relativity. Commenting on the definition, Whitehead says:

> The adequacy of the scheme over every item does not mean adequacy over such items as happen to have been considered. It means that the texture of observed experience, as illustrating the philosophic scheme, is such that all related experience must exhibit the same texture. Thus the philosophic scheme should be 'necessary' in the sense of bearing in itself its own warrant of universality throughout all experience, provided that we confine ourselves to that which communicates with immediate matter of fact [PR, p. 5].

In natural philosophy this "uniform" texture of experience is exemplified as follows: ". . . we can discern in nature a ground of uniformity, of which the more far-reaching example is the uniformity of space-time and the more limited example is what is usually known under the title, The Uniformity of Nature" (R, p. 14). Furthermore, in view of his dissatisfaction with the non-uniformly curved space-time of Einstein's general theory of relativity, Whitehead developed an alternative theory involving only Euclidean, and therefore uniformly curved, space-time.

In the opening pages of R Whitehead gives a succinct statement of what, in his view, philosophy of science should accomplish:

> The philosophy of science is the endeavour to formulate the most general characters of things observed. These sought-for characters are to be no fancy characters of a fairy tale enacted behind the scenes. They must be observed characters of things observed. Nature is what is observed. . . . Thus the philosophy of science only differs from any of the special natural sciences by the fact that it is natural science at the stage before it is convenient to split it up into its various branches [R, p. 5].[3]

The type of philosophy of science which Whitehead is here implicitly rejecting holds that the concepts or entities employed in the formulations of scientific theories are essentially arbitrary or conventional expedients which just "happen" to lead to confirmable laws. Another way of putting Whitehead's view is in

---

[3] Cf. CN, p. 46: "The primary task of a philosophy of natural science is to elucidate the concept of nature, considered as one complex fact for knowledge, to exhibit the fundamental entities and the fundamental relations between entities in terms of which all laws of nature have to be stated, and to secure that the entities and relations thus exhibited are adequate for the expression of all the relations between entities which occur in nature."

terms of his opposition to the bifurcation of nature: Whitehead is contending that there can be no ultimate separation of nature into two realms, one containing the space-time, the electrons, etc., of theoretical physics, the other containing the wealth of perceived qualities which pervade all our experience.

You cannot cling to the idea that we have two sets of experiences of nature, one of primary qualities which belong to the objects perceived, and one of secondary qualities which are the products of our mental excitements. All we know of nature is in the same boat, to sink or swim together. The constructions of science are merely expositions of the characters of things perceived [CN, p. 148].

As a sharp contrast to Whitehead's view one might recall Einstein's vivid remark, that the relation between scientific concepts and sensations "is not like that of the soup to the beef, but rather like that of the cloakroom check number to the coat."[4]

Before entering into details, we might ask what general criteria Whitehead proposes for the construction of an adequate philosophy of science. Throughout his philosophical writings, Whitehead appeals to two fundamentally different types of evidence; for example, one learns that "there are two gauges through which every theory must pass. There is the broad gauge which tests its consonance with the general character of our direct experience, and there is the narrow gauge which is . . . the habitual working gauge of science" (R, pp. 3–4). These two types, or sources, of evidence give to Whitehead's philosophy its peculiar dual aspect, "aesthetic" and "mathematical." Any attempt to evaluate Whitehead's philosophy must ultimately balance claims of the following two kinds: (1) ". . . the ultimate appeal is to naïve experience" (SMW, pp. 129–30); "philosophy is either self-evident, or it is not philosophy"[5] (MT, p. 67); (2) ". . . 'proof' is one of the routes by which self-evidence is often obtained" (MT, p. 66); "Speculative philosophy is the endeavour to frame a coherent, logical, necessary system of general ideas in terms of which every element of our experience can be interpreted" (PR, p. 4).

Perhaps the clearest presentation of the two aspects occurs in Parts III and IV of PR, entitled, respectively, "The Theory of Prehension" and "The Theory

---

[4] A. Einstein, "Physik und Realität," p. 317. Einstein holds that concepts are logically independent of sense-experience and that we can view only as miraculous the fact that our sense-experience can be unified by our freely created concepts; while Whitehead takes as a fundamental desideratum of scientific reasoning the degree to which it is "the necessary outcome of a harmony between thought and sense-presentation" (ASI, p. 218).

The significance of this contrast between Whitehead and Einstein is discussed at some length in my article, "Philosophic Principles and Scientific Theory."

[5] Whitehead continues: "The attempt of any philosophic discourse should be to produce self-evidence. Of course it is impossible to achieve any such aim." This suggests that all of Whitehead's excursions into such *non*-self-evident matters as his definitions of points and straight lines are ultimately for the sake of inducing assent to his basic *aesthetic* categories. This suggestion has the merit of indicating a possible relation between the two types of evidence in Whitehead's philosophy. For, in a philosophy like Whitehead's which stresses the continuity and mutual relevance of categories and subject matters, there can be no ultimately different types of evidence. Of course, this suggestion is a mere hint as to the solution of a very complex problem in Whitehead's philosophy.

of Extension." The theory of prehension is concerned with the genetic character of actual entities, i.e., the process of concrescence whereby an occasion of experience appropriates antecedent data and emerges with its own unique increment of value. The entire discussion is in terms of subjective forms, subjective aims, lures for feeling, and satisfactions—in short, it is an attempt to describe systematically the most general features of experienced aesthetic value. The theory of extension, on the other hand, is a morphological treatment of actual entities. The difference between the two ways of "dividing" an actual entity is that "genetic division is division of the concrescence; coordinate division is division of the concrete" (PR, p. 433). This means that in coordinate division one confronts an achieved, fully determinate, static occasion of experience and examines the *possible* divisions of the region occupied by this occasion and the *possible* relations of these divisions to other regions. In this way one attempts to find the most universal set of extensive (i.e., spatio-temporal) relations which coordinate in a homogeneous way the divisions *within* one actual entity and the divisions *among* all actual entities. The theory of extension is thus essentially mathematical and, in particular, geometrical in character.[6]

In Whitehead's three early works on the philosophy of science, his most significant and characteristic appeal for evidence is to the character of immediate sense-experience. What he is seeking is not so much to invent new scientific theories (although that is involved to a certain extent) as to build more substantial foundations for both old and new theories. Thus, in the Preface to CN Whitehead says that "The object of the present volume and of its predecessor [PNK] is to lay the basis of a natural philosophy which is the necessary presupposition of a reorganised speculative physics" (CN, pp. vii–viii). For example, he does not argue against the utility and theoretical importance of the concepts of points or instantaneous spaces in theoretical physics; rather he attempts to show how these high-level abstractions can be systematically derived from certain aspects of sense-experience.

Whitehead's point of departure in his analysis of the foundations of natural science is a critique of what he calls the "classical concept of the material world." According to this concept,

... the class of ultimate existents is composed of three mutually exclusive classes of entities, namely, *points of space, particles of matter*, and *instants of time*. Corresponding to these classes of entities there exist the sciences of *Geometry*, of *Chronology*, which may be defined as the theory of time considered as a one-dimensional series ordinally similar to the series of real numbers, and of *Dynamics*. There appears to be no science of matter apart from its relations to time and space [MC, p. 467].

[6] For more detailed discussion of the contrast in question see PR, pp. 334 ff., 447–48. In one of his last published writings, Whitehead refers to the need for a *synthesis* of the aesthetic with the mathematical: "When in the distant future the subject [symbolic logic] has expanded, so as to examine patterns depending on connections other than those of space, number, and quantity—when this expansion has occurred, I suggest that Symbolic Logic, that is to say, the symbolic examination of patterns with the use of real variables, will become the foundation of aesthetics" (M, p. 99). Cf. MT, p. 84: "I suggest . . . that the analogy between aesthetics and logic is one of the undeveloped topics of philosophy."

In addition to the three classes of ultimate existents there are three "fundamental relations": the *time-relation* (a dyadic serial relation whose field is the instants of time); the *essential (space-) relation* (a triadic relation of linear ordering whose field is the points of space); and a set of *extraneous relations* equal in number to the number of particles (triadic relations of occupation, the field of any one of them being a single specific particle and all the instants of time and all the points of space).

The classical concept of the material world, however, is open to several serious criticisms:

1. Time and space are absolute, i.e., completely independent of their contents (which corresponds, in the above formulation, to the fact that in addition to the extraneous relations, undefined temporal and spatial relations are also required). But there does not seem to be any observational basis for either the time-relation or the space-relation:

I cannot in my own knowledge find anything corresponding to the bare time of the absolute theory [CN, p. 34].[7]

The knowledge of bare space, as a system of entities known to us in itself and for itself independently of our knowledge of the events in nature, does not seem to correspond to anything in our experience [CN, p. 37].

2. Given the three classes of ultimate existents and the three fundamental relations, one would expect a complete description of the material world to be possible in terms of its successive states, each such state being an instantaneous distribution of matter throughout space. In fact, however, a complete description of the material world requires the introduction of concepts like velocity, acceleration, momentum, and kinetic energy—more generally, the introduction of the notion of an instantaneous *state of change* (this corresponds to the fact that in classical dynamics the physical state of a particle requires the specification not only of the position but also of the velocity—or momentum or kinetic energy—of the particle). Now, the notion of an instantaneous state of change is distinctly difficult and requires explication. A perfectly satisfactory *mathematical* definition of, say, instantaneous velocity, can be given using the limit procedures of the differential calculus, but since these procedures necessarily refer to instants other than the given instant, the mathematical definition only confirms the inadequacy of the classical view that nature can be completely described in terms of its properties at an instant.[8] In sum, the classical view "has ruled out in

[7] Cf. TSM, pp. 45–6: ". . . time as a succession of instants corresponds to nothing which falls within my own direct knowledge. I can only think of it metaphorically either as a succession of dots on a line or as a set of values of an independent variable in certain differential equations. I cannot dissociate time from concrete nature, and then know nature as at an instant of time; nor am I aware of any fact which is instantaneous nature."

[8] Cf. MT, p. 200: "The mathematical subtleties of the differential calculus afford no help for the removal of this difficulty. We can indeed phrase the point at issue in mathematical terms. The Newtonian notion of occupancy corresponds to the value of a function at a selected point. But the Newtonian physics requires solely the limit of the function at that point. And the Newtonian cosmology gives no hint why the bare fact which is the value should be replaced by the reference to other times and places which is the limit."

advance any physical relationships between nature at different instants, and all that is left to connect nature at one instant with nature at another instant is the identity of material and the comparisons of the similarities and differences made by observant minds" (TSM, p. 45).[9]

3. Instants, points, and particles are all dubious entities from the point of view of immediate sense-experience: ". . . the starting point of a discussion on the foundations of geometry is a discussion of the character of the immediate data of perception. It is not now open to mathematicians to assume *sub silentio* that points are among these data" (PNK, p. 5). Similar statements can obviously be made about instants and particles. This criticism may be extended to relational theories of space and time, including the modern theory of relativity in most of its usual formulations.[10]

Whitehead himself adopts a relational theory of space and time, in which the relata are four-dimensional *events*, i.e., occurrences with both a temporal and a spatial spread. All scientific explanation must ultimately be formulated in terms of spatio-temporal relations of such events. Whitehead's procedure enables him to avoid all three of the above difficulties (here I anticipate later detailed discussions). (1) Both time and space are abstractions from events and derive all of their characteristic properties from the perceived properties of events. Furthermore, the sciences of geometry, chronology, and dynamics become significantly interrelated. (2) The notion of a state of change is no longer defined in terms of the properties of particles at different instants but is taken instead

[9] Biology fares even worse on the classical view: "Every expression of life takes time. Nothing that is characteristic of life can manifest itself at an instant. Murder is a prerequisite for the absorption of biology into physics as expressed in these traditional concepts" (TSM, p. 45).

[10] In particular, this criticism applies to each of the four concepts of the material world analyzed by Whitehead in MC as alternatives to the classical concept, because (if for no other reason) each of these four concepts takes instants as one class of ultimate existents. Also, the really original portion of MC (the analysis of Concepts IV and V) has as one of its central motivations the definition of points in terms of "lines" which may be identified with "the lines of force of the modern physicist" (MC, p. 482)—hardly a useful result from the point of view of Whitehead's natural philosophy. However, Whitehead explicitly disavows any direct philosophic purpose in his study of possible concepts of the material world: "The general problem is here discussed purely for the sake of its logical (*i.e.*, mathematical) interest. It has an indirect bearing on philosophy by disentangling the essentials of the idea of a material world from the accidents of one particular concept" (MC, p. 465). Thus Whitehead's results in this early monograph do not constitute a significant anticipation of his later philosophy of science—a subject to which those results make at best an indirect contribution. Does, or did, the monograph in question have any bearing on theoretical physics? Since it has been alleged that Whitehead's results anticipated Einstein's general theory of relativity, I think it important to emphasize that the answer here too is clearly negative. Wisely enough, Whitehead himself makes no such claims (granting the possibly ironic overtones of the word "existing" in the following passage): "The problem might, in the future, have a direct bearing upon physical science if a concept widely different from the prevailing concept could be elaborated, which allowed of a simpler enunciation of physical laws. But in physical research so much depends upon a trained imaginative intuition, that it seems most unlikely that existing physicists would, in general, gain any advantage from deserting familiar habits of thought" [MC, pp. 465–66].

as one of the ultimate notions of natural philosophy. (Whitehead admits, however, in Note II to the second edition of PNK that even he failed to "insist properly" on the primacy of this notion: ". . . the true doctrine, that 'process' is the fundamental idea, was not in my mind with sufficient emphasis" [PNK, p. 202].) (3) By means of the method of extensive abstraction instants, points, and particles are defined in terms of "the immediate data of perception."

In order to understand and to evaluate the above achievements by Whitehead one must first study some important preliminary ideas stemming from the multiplicity of space-time systems inherent in Einstein's special theory of relativity.

CHAPTER II

EINSTEIN'S

SPECIAL THEORY

OF RELATIVITY

1. Space and Time in Special Relativity

In his special theory of relativity Einstein explicitly rejects the conceptions of absolute space and absolute time which Newton had introduced in his *Principia*. Not only does Einstein claim that there is no observational basis for the assumption of absolute space and time, but, pointing to the null results of the various ether-drift experiments, he shows that it is no longer tenable to believe in the existence of a stationary ether, with respect to which absolute motion can be conceived as taking place.

The principal features of Einstein's special theory of relativity which are relevant to Whitehead's philosophy of science may be summarized as follows.[1]

1. It is assumed that space is measured by means of a standard rigid rod and time by means of a standard clock (such as a periodic mechanical system, e.g., a pendulum). Physical space is to be thought of as all possible quasi-rigid extensions of the standard rigid rod; measurements with such rods lead to the result that physical space is three-dimensional and Euclidean. Physical time is to be thought of as the totality of readings of the standard clock; times of events not occurring in the immediate vicinity of the clock are to be determined with the aid of signals of extremely high velocity sent from the events to the clock. An "inertial system" may now be defined as a set of spatial and temporal co-ordinates with respect to which the laws of classical mechanics hold good to a first approximation, i.e., for bodies moving with sufficiently small velocities.[2]

---

[1] An elementary exposition of the special theory of relativity can be found in A. Einstein, *Relativity, the Special and General Theory*. For further details, one can consult Einstein's original paper of 1905 "On the Electrodynamics of Moving Bodies" in H.A. Lorentz *et al.*, *The Principle of Relativity;* also the following more advanced treatises may be recommended: A. Einstein, *The Meaning of Relativity;* H. Weyl, *Space-Time-Matter;* P. Bergmann, *Introduction to the Theory of Relativity;* L. Silberstein, *The Theory of Relativity;* C. Møller, *The Theory of Relativity;* W. Pauli, *Theory of Relativity*.

[2] The qualification "to a first approximation" avoids what might otherwise seem a logical circularity in Einstein's procedure. For, while it is true that one could test the validity of

In order to obtain laws of mechanics which are true in higher approximation, one must first of all introduce a precise meaning for the synchronism of different clocks. Since indefinitely fast physical signals are simply not available, one proceeds by stipulating that light-rays—the fastest physical signals known— will move with equal speeds in either direction between two fixed points in an inertial system. Clock-synchronism is then defined as follows: two clocks at rest within a given inertial system are synchronous if a light-signal sent from the first clock arrives (and is reflected back) at the second clock at a time (as determined by the second clock) midway between the times of departure and return of the light-signal (as determined by the first clock). If A and B are the positions of two clocks within a given inertial system K, this definition of clock-synchronism may be expressed as follows:

$$t_B - t_A = t'_A - t_B , \quad \text{or} \quad t_B = \frac{t_A + t'_A}{2},$$

where t_A is the time of departure of the light-signal from A (by the clock at A), t_B is the time of arrival and reflection of the light-signal at B (by the clock at B), and t'_A is the time of return of the light-signal to A (by the clock at A) It is obvious that this definition of clock-synchronism amounts to a criterion for the simultaneity of distant events. Einstein assumes that the definition may be consistently extended to any number of points within K, and that the relation of synchronism so defined is symmetrical and transitive. Furthermore, it is to be understood that in any other inertial system K', space and time are to be measured using procedures identical to those in K and by means of rigid rods and clocks at rest in K' but otherwise identical with those in K. Finally, Einstein assumes on the basis of experiment that the speed of the light-signals moving between A and B (a distance AB) is a universal constant, $c = (2AB)/(t'_A - t_A)$, which represents the numerical value of the speed of light in empty space in all inertial systems.

2. Two postulates are fundamental to the theory: (*a*) *the special principle of relativity*, which states that all inertial systems are equivalent for the description of natural phenomena, i.e., any laws of nature must have the same form for all coordinate-systems in which the equations of Newtonian mechanics hold (this follows as an inductive generalization from the observed relativity of electrodynamic phenomena and from the null results of the Michelson-Morley, and other, ether-drift experiments); (*b*) *the principle of the constant velocity of light*, which states that light is always propagated in empty space (relative to a given inertial system) with a fixed velocity c, independent of the state of motion of the source of light (this follows from the validity of the Maxwell-Hertz electrodynamic theory for empty space).

---

classical mechanics only if one already possessed instruments for measuring space and time, one would *not* need clocks synchronized according to Einstein's light-signal definition in making such a test. To determine the validity of classical mechanics to a first approximation involves precisely the assumption that physical signals of extremely high velocity are available (and that light, for example, is one of them).

3. From the fundamental postulates and definitions of the theory, one can derive, by purely mathematical procedures, a set of equations (called the "Lorentz-Einstein equations") which specify how the results of space and time measurements made in different (inertial) coordinate-systems are related to one another. These equations constitute the heart of the special theory of relativity. In order to formulate the equations, let us assume that x, y, z and x', y', z' represent the three rectangular Cartesian spatial coordinates of the inertial systems K and K' respectively; and t and t' represent the temporal coordinates of K and K' respectively. Let us assume further that the origins and respective spatial axes of K and K' coincide at the time $t = t' = 0$ and that thereafter K' moves with a uniform velocity v, relative to K, along the coincident x- and x'-axes. Finally, let c represent the universal constant whose value is the speed of light in empty space. The Lorentz-Einstein equations then have the following form:

$$\left. \begin{aligned} x' &= \frac{x - vt}{(1 - v^2/c^2)^{1/2}}, \\ y' &= y, \\ z' &= z, \\ t' &= \frac{t - vx/c^2}{(1 - v^2/c^2)^{1/2}}. \end{aligned} \right\} \quad (1)$$

We note immediately that for values of v much smaller than c, equations (1) become practically equivalent to the Galilean transformation equations of classical mechanics:

$$\left. \begin{aligned} x' &= x - vt, \\ y' &= y, \\ z' &= z, \\ t' &= t. \end{aligned} \right\} \quad (2)$$

Also, upon solving the equations (1) for x, y, z, and t in terms of x', y', z', and t', one obtains a set of equations identical in form with equations (1), except that the sign of v is reversed. The obvious interpretation of this result is that K is moving with the velocity $-v$, relative to K', along the coincident x- and x'-axes. Thus the relative velocities of K and K' are equal and opposite—not a trivial conclusion in view of the fact that there is no method for directly comparing the space-units and time-units adopted in K with those adopted in K'.

Some of the more important consequences of the Lorentz-Einstein equations will now be listed.

4. Associated with any fixed location in K there is a unique order of events occurring *at* this particular location; in other words, time-measurements made in other inertial systems will agree on this order. In this sense there is an absolute *local* time for every point of an inertial system.

5. Furthermore, for inertial systems whose coordinate origins momentarily coincide (such as K and K'), a unique non-overlapping classification of all events into past, future, and indeterminate (neither-past-nor-future) is possible at the moment of coincidence. Even at that moment, however, there will be different meanings for simultaneity in the inertial systems in question.

6. There is not, in general, an invariant temporal lapse between any two events; nor is there, in general, an invariant distance between any two simultaneous events. In other words, observations in different inertial systems will generally yield varying times between a pair of events and varying distances between a pair of simultaneous events. These varying times and distances are by no means arbitrary or unpredictable but are rather, as we shall see, a definite function of the relative motions of the inertial systems. (This relativity of space and this relativity of time are each, of course, incompatible with the concepts of space and time in classical mechanics.)[3] More specifically, the following kinematic relationships hold. (a) For an inertial system moving (uniformly) relative to a clock, its rate is slower than for an inertial system at rest relative to the clock, the "retardation" being given by the factor $1/(1 - v^2/c^2)^{1/2}$. (b) The length of a rigid rod measured at right angles to the (uniform) relative motion of two different inertial systems is the same in either inertial system; the length of a rigid rod measured in any other direction will be less (by an amount dependent upon the magnitude and direction of v) for the inertial system in motion relative to the rod than for the inertial system at rest relative to the rod, the maximum "contraction" occurring for measurements parallel to the relative motion of the two inertial systems and being given by the factor $(1 - v^2/c^2)^{1/2}$.

7. There is a measurable quantity defined for any pair of events and called the *interval* between those events, which has the same value for all inertial systems. The interval may be interpreted as ordinary spatial distance for simultaneous events and as ordinary temporal duration for events at the same place. Algebraically expressed, the differential of the interval squared, ds^2, is given by:

$$d s^2 = -(dx^2 + dy^2 + dz^2) + c^2 dt^2 , \qquad (3)$$

---

[3] In a certain sense, the relativity of time is more fundamental: if time-lapses between events are relative, then the meaning of simultaneity (the special case of zero time-lapse) is relative, from which it follows that the distance between simultaneous events must also be relative. Thus the relativity of time is often singled out as the most significant philosophical implication of the theory of relativity. K. Gödel, for example, puts it this way: "The existence of an objective lapse of time . . . means (or, at least, is equivalent to the fact) that reality consists of an infinity of layers of 'now' which come into existence successively. But, if simultaneity is something relative . . . , reality cannot be split up into such layers in an objectively determined way. Each observer has his own set of 'nows,' and none of these various systems of layers can claim the prerogative of representing the objective lapse of time" ("A Remark about the Relationship between Relativity Theory and Idealistic Philosophy," *Albert Einstein: Philosopher-Scientist*, ed. P. A. Schilpp, p. 558). Cf. Whitehead (CN, p. 72): ". . . there is no unique factor in nature which for every percipient is pre-eminently and necessarily the present."

where dx, dy, dz are the respective infinitesimal increments of the three rectangular Cartesian spatial coordinates, dt is the infinitesimal increment of the time coordinate, and c is the velocity of light. Using the Lorentz-Einstein equations one can easily compute the interval for K and for K'; the result, namely, $ds^2 = ds'^2$, proves the above assertion about the invariance of interval. The interval is called "timelike" when $ds^2 > 0$, "spacelike" when $ds^2 < 0$, and "null" when $ds^2 = 0$. Timelike intervals represent possible paths for material particles, spacelike intervals represent impossible paths for material particles, and null intervals represent possible paths for light-rays.

8. Also deducible from the Lorentz-Einstein equations is the *relativistic addition of velocities theorem*. This theorem provides an answer to questions of the following kind: suppose a body is moving, relative to K, with velocity u along the x-axis; what is the velocity of the body relative to K'? The answer of classical mechanics is given by the Galilean addition of velocities theorem:

$$u' = u - v \,, \tag{4}$$

which may easily be generalized to cases in which u is not parallel to v. The formula corresponding to equation (4) in relativistic mechanics is:

$$u' = \frac{u - v}{1 - u\,v/c^2} \,, \tag{5}$$

which again may be generalized to cases in which u is not parallel to v. It follows from equation (5) that the addition of any two velocities less than or equal to c yields a velocity less than or equal to c, and that, as implicitly assumed in Einstein's postulates, no material body can exceed the velocity c.

9. All of the above kinematic results of the special theory of relativity have have now been amply confirmed, some of them, however, only quite recently. The contraction of rigid rods was, of course, confirmed by the Michelson-Morley and numerous subsequent ether-drift experiments; the retardation of clocks (or, as it is sometimes called, the "dilatation of time") was first confirmed by the experiments of Ives and Stilwell (1938) on the transverse Doppler effect, and later by the data on meson disintegration rates observed by Rossi and Hall (1941); the application of the Lorentz-Einstein equations to the behavior of light leads to a formula for aberration which agrees with all known observations (here the second-order effect is far below the limits of experimental accuracy) and to a formula for optical "dragging phenomena" which agrees with the results of Fizeau as well as the later and more precise results of Zeeman (1914–15).[4]

10. Guided by the form of the Lorentz-Einstein equations, the classical concepts of momentum and energy can be redefined and a system of relativistic mechanics developed (see below, p. 177). If it is assumed that conservation of momentum is (as in classical mechanics) universally valid, it turns out that the

[4] For an account of the more recent experiments see Møller, *The Theory of Relativity*, pp. 62–66.

two independent classical conservation laws for mass and energy must be replaced by a single conservation law for energy-plus-mass. This last result is usually referred to as the "equivalence" of mass and energy as expressed in the equation:

$$E = m c^2 = \frac{m_0 c^2}{(1 - v^2/c^2)^{1/2}},$$

(6)

where E is the amount of energy corresponding to the inertial mass m, c is the velocity of light, m_0 is the so-called rest-mass of m, and v is the velocity of m. Confirmation of equation (6) came only with the experimental work on atomic nuclei beginning in the 1930's.[5]

2. Minkowski's Space-Time Geometry

The important thing for our immediate purposes about the results summarized above is that they imply the possibility of many space-time systems in a way not implied by the classical conceptions of space and time. (This multiplicity of space-time systems is expressed mathematically in the Lorentz-Einstein equations by the inextricable interrelation of the space and time variables.) The import of this multiplicity of space-time systems can be clearly illustrated by using a geometrical representation first proposed by Minkowski.[6] It should be pointed out, however, that the formulations to be presented are not always accepted by Whitehead (e.g., he is highly critical of the central role accorded the velocity of light in Einstein's theory). On the other hand, Whitehead has often acknowledged his great debts to Einstein and Minkowski, and there can be little doubt that he was especially influenced by the latter's four-dimensional geometry (as will become evident below).

For purposes of comparison with relativity kinematics, let us begin with the geometrical representation of classical kinematics. Space is assumed to be three-dimensional and Euclidean. In such a space, points can be conveniently described by means of rectangular Cartesian coordinates, i.e., by three numbers representing the respective distances of any point from three mutually perpendicular coordinate axes. In classical mechanics time constitutes a one-dimensional continuum entirely independent of space; it is therefore quite arbitrary how we decide to orient the time-axis in our geometrical diagrams for describing motion. In Figure 1 the time-axis is taken, for convenience, as perpendicular to the space-axis; also for convenience, and without real loss of generality, two space-axes have been omitted. A point in Figure 1 represents an instantaneous and spatially infinitesimal occurrence, sometimes referred to as a "point-event," but here in the interest of brevity called simply an "event." (In Minkowski's terminology, a point in four-dimensional space-time is called a *world-point* and a line in four-dimensional space-time is called a *world-line*.)

[5] For a summary account see *ibid.*, pp. 85–91.
[6] H. Minkowski, "Space and Time" in Lorentz *et al.*, *The Principle of Relativity*.

Consider (Figure 1) an inertial system K with coordinate axes x, t and origin O. Positions of a body at rest at the origin of the x-space are represented by the points on the t-axis (i.e., by a series of successive events at the same place). The x-axis represents a set of events simultaneous with the event represented by O, while straight lines parallel to the x-axis would represent earlier or later sets of simultaneous events. Any straight line in the x-t coordinate-system, with slope between zero and infinity, represents a possible path of a body moving with (finite or infinite) constant speed; thus, for example, the line OO' represents a body moving with velocity $v = x/t$, if (x, t) are the coordinates of O'. (In our simplified diagram, changes of direction are obviously impossible because there is just a single spatial axis; nothing essential, however, is lost by this simplifica-

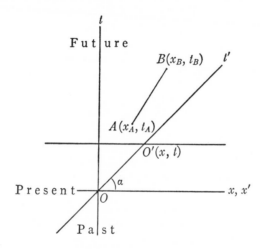

FIG. 1.—Classical kinematics

tion.) Let us now imagine a second inertial system K' moving along the path OO' (with respect to K). At a certain instant the origins of K and K' will coincide. Because of the absoluteness of time in classical mechanics, K and K' will have identical meanings for "the present" at the instant of coincidence, and hence will adopt the same space-axis x at that instant. At any time after the instant of coincidence of K and K' their space-axes will coincide, but their origins will gradually separate with the relative speed v. According to the principle of relativity of classical mechanics, instead of regarding K' as moving relative to K, K may with equal right be regarded as moving relative to K'. Thus, the time-axis t' of K' will be represented by the line OO' inclined to the x-axis at an angle $\tan a = 1/v$ (i.e., by a series of successive events at the same place). Also, t' may be considered perpendicular to x' even though it is impossible to represent this in a diagram in which x is already perpendicular to t. Since time is assumed to be absolute, time measurements in K and K' must agree; hence the scales for plotting times along the t- and t'-axes must differ (in fact the scale along t' will be greater than that along t in the ratio $1/\sin \alpha = [1 + v^2]^{1/2}$.

The question now arises: how will the spatial and temporal coordinates assigned to a given event by K and K', respectively, be related to one another? Classical mechanics answers this question by means of the Galilean transformation equations which we have already encountered above (equations [2]). For the event A (Figure 1) these equations give:

$$x'_A = x_A - vt_A ,$$
$$t'_A = t_A ,$$
$$\left.\vphantom{\begin{matrix}x'_A = x_A - vt_A \\ t'_A = t_A\end{matrix}}\right\} \quad (7)$$

where (x_A, t_A) and (x'_A, t'_A) are, respectively, the coordinates assigned by K and K' to the event A, and v is the velocity of K' as measured by K. We can now find how the time-lapse between any pair of events, as determined by K, is related to the time-lapse between those same events as determined by K'; and similarly for the distance between any pair of events. Let the two events be represented by A and B (Figure 1); then we have:

$$x'_B - x'_A = x_B - x_A - v \, (t_B - t_A) , \qquad (8)$$
$$t'_B - t'_A = t_B - t_A . \qquad (9)$$

Equation (9) expresses, of course, the absoluteness of time. But the distance between the two events is *not* the same for K and K'; only if one considers the distance between *simultaneous events* (for which $t_A = t_B$ and $t'_A = t'_B$) is the distance the same for K and K' $(x'_B - x'_A = x_B - x_A)$.

Our results for classical kinematics may be summed up by saying that there are two classical kinematic invariants: the time-lapse between any pair of events and the distance between any pair of simultaneous events. Incidentally, this accounts for the independence of the space and time continua in classical kinematics: what is required for a genuine space-time continuum is a concept of spatio-temporal "distance" (i.e., an expression for a space-time metric) which is invariant for all inertial systems, and no such concept is definable in classical kinematics (since Δt^2 is an invariant while Δx^2 is not, it follows that neither $\Delta x^2 + \Delta t^2$ nor $\Delta x^2 - \Delta t^2$—the two plausible candidates for a space-time metric —is an invariant). Thus the line-segment AB in Figure 1 has no physical significance in the sense that no single numerical value can be associated with it for all inertial systems.

Figure 1 has been drawn from the point of view of the inertial system K. From the point of view of K' an analogous diagram could be constructed, interchanging the roles of K and K', with the sole difference that K would be depicted as moving with a constant velocity $-v$ relative to K'. The path of K (and hence the t-axis) would be represented in the new diagram by a straight line inclined to the common x- and x'-axis at an angle equal to $\pi - \alpha$. This symmetry between the two inertial systems is preserved in special relativity kinematics, but its geometrical representation is, as we shall see, somewhat more complicated.

We turn now to special relativity kinematics (see Figure 2). Space is assumed to be three-dimensional and Euclidean, but now, in addition, space-time is assumed to be "pseudo-Euclidean" (which will be explained in a moment). The possibility of introducing a metric for space-time in special relativity mechanics arises from the existence within the special theory of relativity of an invariant "interval" between any pair of events. By means of a four-dimensional rec-

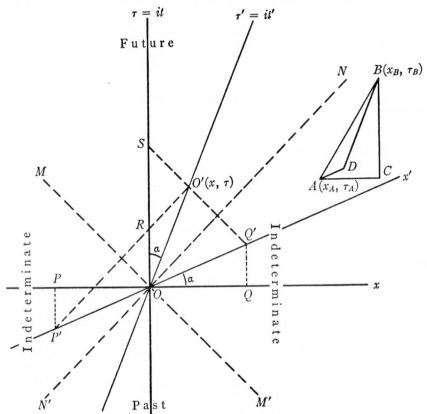

FIG. 2.—Special relativity kinematics

tangular Cartesian coordinate-system (in which three axes are spatial and one is temporal) the metric for space-time can be expressed in the simple form given by equation (3). Such a metric is said to be "pseudo-Euclidean" because of the disagreement in sign between the three spatial increments (all negative) and the temporal increment (positive) in the formula for interval. By using imaginary time, $\tau = \sqrt{-1}t = it$, instead of the real time t, as the fourth coordinate, one obtains an expression for ds^2 symmetrical in the four coordinates:[7]

$$d s^2 = - (d x^2 + d y^2 + d z^2 + c^2 d \tau^2) . \qquad (10)$$

[7] This idea was first suggested by Minkowski (*ibid.*). The metric, of course, remains pseudo-Euclidean, as is shown by the fact that ds^2 may vanish even when none of the four differentials, dx^2, dy^2, dz^2, $d\tau^2$, vanishes.

A further simplification results if we assume that $c = 1$, i.e., that units for space and time are so chosen that the velocity of light equals unity. It should be noted that the adoption of equation (10) as the expression for space-time metric makes it imperative—and no longer, as in classical kinematics, a mere convention—to draw the τ-axis perpendicular to the x-, y-, and z-axes. Thus, in Figure 2 (where, as in Figure 1, two space-axes have been suppressed, and a point represents an event) the τ-axis is necessarily perpendicular to the x-axis.

According to the special theory of relativity, as already noted, no material body can move with a speed exceeding c, the speed of light. We represent this in Figure 2 as follows. First we draw a pair of broken lines, MM' and NN', bisecting the four quadrants in the x-τ coordinate-system. The lines MM' and NN', and any other lines parallel to either of them, represent possible paths of light-rays (note that the slope of the lines is unity, as it should be since we have assumed that $c = 1$). We now assume that no material body can move along a curve, in the x-τ coordinate-system, whose slope at any point is less than unity in absolute value. Straight lines with slope greater than unity in absolute value (i.e., straight lines which lie wholly in the regions bounded by MON and $M'ON'$) will represent possible paths (or world-lines) of bodies moving with (necessarily finite) constant speed; thus, for example, the line OO' represents a body moving with velocity $v = x/t$ (relative to K), if $(x, \tau = it)$ are the coordinates of O'.

Consider (Figure 2) an inertial system K with coordinate axes x, τ and origin O. Positions of a body at rest at the origin of the x-space are represented by the points on the τ-axis (i.e., by a series of successive events at the same place). The x-axis represents a set of events *simultaneous for K* with the event represented by O, while straight lines parallel to the x-axis would represent earlier or later sets of events *simultaneous for K*. Let us imagine a second inertial system K' moving along the path OO' (relative to K). The slope of OO' is $\tau/x = -1/iv$, so that $\tan a = -iv$. At a certain instant the origins of K and K' will coincide. Now, according to special relativity kinematics (in particular, the Lorentz-Einstein equations) K and K' will have entirely different sets of coordinate axes. In fact, it is not difficult to prove that, *from the point of view of K, K'* will be using the x'-τ' coordinate axes (see Figure 2), which are oriented toward the x-τ axes at an angle a such that $\tan a = -iv$. To see this, we may reason as follows. First of all, because of the principle of relativity, we are entitled to think of K' at rest instead of K; thus, the points along OO' will represent the successive positions of the origin of K' at different times; in short, OO' will represent the (imaginary) time-axis of K' (call it τ'). As for the space-axis, x', it must be chosen so that the speed of light in K' turns out to be the universal constant, $c = 1$; from which it follows that the x'-axis must make the same angle with NN' that the τ'-axis makes with NN'. Also, since the angle between τ and τ' is a, the angle between x and x' must be a. That the angle between the x'- and

τ'-axes will be a right angle for K' follows from the assumption that the metric of space-time is pseudo-Euclidean for K' just as for K.[8]

It is now readily seen that the lines MM' and NN' divide all possible events into three mutually exclusive classes (relative to an event at O): past (the region bounded by $N'OM'$), future (the region bounded by MON), and indeterminate (the regions bounded by MON' and NOM'). The events in the last category are unusual (from the point of view of classical mechanics) in that, although all of them are indeterminate with respect to O, they are not necessarily indeterminate with respect to one another. Using Whitehead's term "co-present" to refer to the relation which subsists between each pair of members of the class of indeterminate events, we note that this relation of co-presence is symmetrical but non-transitive. To see this, consider first the events P and Q along the x-axis (for the sake of a later example P and Q are chosen equidistant from the origin). Here, P is co-present with O and O is co-present with Q, and also P is co-present with Q; hence, co-presence is not intransitive. Now, let P' and Q' represent the projections of P and Q, respectively, on the x'-axis. Consider the events P, P', and O. Here, P is co-present with O and O is co-present with P', but P is not co-present with P' (in fact, P is future with respect to P'); hence, co-presence is not transitive. Thus, being neither transitive nor intransitive, co-presence is non-transitive.

Of the events co-present with O, different sets will, of course, be simultaneous for different inertial systems. Thus, P and Q are simultaneous for K but not for K', while P' and Q' are simultaneous for K' but not for K. The relationships between P' and Q' are confirmed by the fact that light-signals leaving P' and Q' will reach the τ'-axis at the same instant (at O') but will reach the τ-axis at different instants (at R and S). In accordance, then, with Einstein's definition of simultaneity (in its appropriate formulations for K and K', respectively), P' and Q' are simultaneous for K', while P' occurs before Q' for K.

Suppose now that P and Q represent, at the instant $\tau = 0$, the ends of a rigid rod at rest in K. The length of the rod in K is given simply by the line-segment PQ. What will be the length of the rod in K'? The rod is moving with respect to K', so that one must locate a pair of points representing the positions of the ends of the rod *at the same time* (according to the definition of simultaneity in K'). But we have already found such a pair of points in P' and Q'. To compare the magnitudes of PQ and $P'Q'$, we make use of the pseudo-Euclidean character of the geometrical relations represented in Figure 2. Since OPP' is a right triangle,

$$(P'P)^2 + (PO)^2 = (P'O)^2 .$$

[8] Of course the x'- and τ'-axes are not at right angles in the (Euclidean) plane of the diagram in Figure 2. This suggests an extremely important general precaution to be observed in using Minkowski space-time diagrams: one must be careful to distinguish the geometrical properties *of* the diagram from the geometrical properties (of space-time) being represented *by* the diagram.

But $P'P/PO = \tan \alpha = -iv;$ hence we have

$$(PO)^2 (1 - v^2) = (P'O)^2 ,$$

which is simply a special case of our earlier expression for the contraction of a rigid rod (remembering that here $c^2 = 1$). In precisely similar fashion one can derive the expression for clock retardation.[9] In general, the special relativity transformation equations for distances and (imaginary) time-lapses between any pair of events A and B, as one goes form K to K', are as follows (see Figure 2):

$$x'_B - x'_A = \frac{(x_B - x_A) - v/i \,(\tau_B - \tau_A)}{(1 - v^2)^{1/2}} ,$$

$$\tau'_B - \tau'_A = \frac{(\tau_B - \tau_A) - vi \,(x_B - x_A)}{(1 - v^2)^{1/2}} .$$

These equations (which follow immediately from equations [1]) should be compared with equations (8) and (9). It can easily be seen that neither time-lapse between events nor distance between events is invariant; interval, on the other hand, *is* invariant, as can easily be proved by direct computation. Thus the line-segment AB in Figure 2 has genuine physical significance (unlike the analogous line-segment AB in Figure 1) in that it represents a definite physical magnitude with the same numerical value for both K and K' (as well as for all other co-ordinate-systems moving uniformly relative to K and K'). The line-segment AB may be resolved into "spatial" and "temporal" components in an infinite number of different ways, each way corresponding to a different inertial system (e.g., AC and BC correspond to K, AD and BD correspond to K'). Our results for relativistic kinematics may be summed up by saying that there is a single kinematic invariant in special relativity: the interval between any pair of events.

Figure 2 has been drawn from the point of view of the inertial system K. From the point of view of K' an analogous diagram could be constructed, interchanging the roles of K and K', with the sole difference that K would be depicted as moving with a constant velocity $-v$ relative to K'. The motion of K would be represented in the new diagram by a straight line making an angle α, measured counterclockwise, with the τ'-axis.

[9] The actual values for distances or time-lapses obtained by K and K' respectively are not immediately comparable as they are read off along the x-τ and x'-τ' axes of Figure 2. One must take account of the fact that the *units* of space and time employed by K and K' respectively are not themselves invariant for the two observers. Once more we notice that the geometrical relationships depicted *by* Figure 2 are often quite different from the geometrical relationships *in* Figure 2: the length of $P'O$ in the diagram is greater than the length of PO in the diagram, whereas $P'O$ as measured by K' would actually be less than PO as measured by K. These geometrical relationships are clearly described in chap. iii of A. S. Eddington's *Space, Time, and Gravitation*, with the aid of some well-drawn diagrams. An especially full and careful presentation of Minkowski space-time geometry may be found in J. L. Synge, *Relativity: The Special Theory.*

CHAPTER III

WHITEHEAD'S

PHILOSOPHY

OF NATURE

The opening chapter of PNK is entitled "Meaning," and I believe that careful reading should convince one that the guiding principle in all Whitehead's writings on science (and to a great extent on other subjects as well) is the *analysis of meaning*, in the broadest sense of that phrase. In the chapter referred to Whitehead criticizes traditional physical theories because they can give "no intelligible account of the meaning" of important physical concepts (such as velocity, momentum, and stress). Here, then, I suggest, is the controlling impulse in Whitehead's philosophy of science: the attempt to give an account of scientific concepts and laws which possesses an *intelligible meaning*. Once this is grasped, many things in Whitehead's work which are otherwise obscure become clarified.

There are at least two kinds of explication of the meaning of fundamental physical concepts (such as space and time) which Whitehead explicitly rejects: the first would identify these concepts with certain variables occurring in the mathematical equations used to express physical laws; the second would identify these concepts with certain measuring procedures (in rejecting this view Whitehead is, of course, rejecting the so-called "operationalist" theory of meaning).[1] Whitehead's objections to these two modes of explicating physical concepts are the following:

Time and space are among the fundamental physical facts yielded by our knowledge of the external world. We cannot rest content with any theory of them which simply takes mathematical equations involving four variables (x, y, z, t) and interprets (x, y, z) as space coordinates and t as a measure of time, merely on the ground that some physical law is thereby expressed. This is not an interpretation of what we *mean* by space and time. What we mean are physical facts expressible in terms of immediate perceptions; and it is incumbent on us to produce the perceptions of those facts as the meanings of our terms [PNK, pp. 45–46].

[1] Cf. CN, pp. 120–21; and PR, p. 501: ". . . although 'coincidence' is used as a *test* of congruence it is not the *meaning* of congruence."

... complete accuracy is never obtained [in judgments of spatial and temporal co-incidence], and the ideal of accuracy shows that the meaning [of spatial and temporal congruence] is not derived from the measurement [PNK, p. 56].

Another example of the first mode of explication is the frequently proposed definition of force as the product of mass and acceleration; another example of the second mode of explication is Einstein's definition of simultaneity in terms of light-signals. Whitehead's critique of the latter will be discussed in some detail below (pp. 34 ff.). As to the former, the difficulty with defining force, F, as mass, m, times acceleration, a, is that Newton's second law of motion, $F = ma$, then becomes $ma = ma$, and "It is not easy to understand how an important science can issue from such premisses" (PNK, p. 19).[2]

Whitehead's repeated statements to the effect that philosophy of science must concern itself with "observed characters of things observed" set a problem rather more than they imply any definite philosophical doctrine. The problem can be stated in the form of two questions: (1) What things are in fact observed, i.e., what are the fundamental observed data on which natural science is based? (2) What is an observed character, i.e., what properties not themselves directly observed may legitimately be attributed to observed things? To the first question Whitehead replies not by giving an exhaustive list of the kinds of observed things or a formal definition of "observable" but rather by indicating those features of the observed which he finds fundamental for natural science. These features are called the *natural elements*. The second question is answered by the adoption of the *method of extensive abstraction* as a means for deriving various types of scientific entities from the natural elements.

Detailed discussion of the natural elements is best preceded by a sketch of Whitehead's general conceptions of *nature* and of *sense-perception*.[3] First of all, we note that sense-perception is not the only activity of the mind and hence that "Nature"—the terminus of sense-perception—"is an abstraction from something more concrete than itself which must also include imagination, thought, and emotion" (R, p. 63). Such questions as whether sense-perception

---

[2] Whitehead's own account of the logical status of Newton's second law of motion is very briefly outlined in PNK (pp. 18–19). A fuller account along the same lines (I believe) may be found in W. H. Macaulay, "Newton's Theory of Kinetics." For what it is worth, it may be mentioned that Whitehead was familiar with Macaulay's article, since he cites it in MC (p. 481).

[3] The sketch which follows is based upon PNK, CN, and R. Since each of these books has its own distinctive terminological (and other) features, my discussion is not offered as an adequate account of any one of the books, far less of all of them; instead, I have endeavored to extract the fundamental distinctions and assumptions which the three books share. I should add that each book constitutes a self-contained and closely argued whole which will amply reward careful analysis.

For a wonderfully sympathetic and lucid account of Whitehead's natural philosophy, see C. I. Lewis, "Whitehead and the Categories of Natural Knowledge," *The Philosophy of Alfred North Whitehead*, ed. P. A. Schilpp. Useful introductions to PNK and R are provided by C. D. Broad's critical notices of these books in *Mind*. Also, TSM gives an excellent summary of the non-technical portions of PNK.

always involves thought or whether we can be aware through sense-perception alone of moral and aesthetic qualities do not concern Whitehead: it is sufficient for his purposes to concentrate attention upon what *is* indubitably given in sense-perception. And this indubitably given is the undifferentiated terminus of a process which may be called sense-awareness; this terminus is "the whole occurrence of nature. It is nature as an event present for sense-awareness, and essentially passing. . . . Thus the ultimate fact for sense-awareness is an event" (CN, p. 14). As soon as we reflect upon the content of sense-awareness we automatically begin to differentiate (or "diversify") that content; the differentiated termini of sense-awareness which thought discovers are called *factors*. These same factors, insofar as they function as termini of *thought*, are called *entities*. There are two principal, and, in a way, opposed, types of factors (or entities) which are important in natural science: *events* and *objects*.[4]

An *event* is an actual happening or occurrence in nature; it is "the most concrete fact capable of separate discrimination" (CN, p. 189).[5] "An event is what it is, when it is, and where it is. Externality and extension are the marks of events . . ." (PNK, p. 62). An event cannot change once it has occurred; it can only pass into some other event which includes it. Since events are extended (in several dimensions) an event has parts; and any part of an event is itself an event. Thus, nature may be thought of as a continuous passage of events into other events. Consider an example. Suppose I gaze at a particular patch of red color on the cover of a book in front of me for one second. The concrete situation of that red patch during that second is an event (e_1). Also, the concrete situation of that red patch during some fraction of that same second is another event (e_2), and any (spatial) part of that same situation which lasts the entire second is still another event (e_3). (The two latter events, e_2 and e_3, are "included in," or "extended over by," the first event, e_1.) This example should not be taken to imply that events are necessarily "small" or "brief": "An event does not in any way imply rapid change; the endurance of a block of marble is an event" (R, p. 21). In fact, as we shall see, Whitehead is very much concerned with infinitely extended events.

Objects, unlike events, may recur; in fact, the permanences found in nature

---

[4] Cf. Whitehead's later distinction in PR, p. 33: "Among [the] . . . categories of existence, actual entities and eternal objects stand out with a certain extreme finality. The other types of existence have a certain intermediate character." "Actual entities" (or "actual occasions") are closely related to the earlier "events," "eternal objects" to the earlier "objects." The determination of the exact relationships between these two pairs of categories is a difficult problem. Cf. below, chap. vi, nn. 8, 9, 15; chap. vii, n. 3.

[5] It is important to realize that even events—the most concrete type of factor in nature—presuppose a background or context or totality within which they have been discriminated, namely, *fact*. "Fact enters consciousness in a way peculiar to itself. It is not the sum of factors; it is rather the concreteness (or, embeddedness) of factors, and the concreteness of an inexhaustible relatedness among inexhaustible relata" (R, p. 15). Sometimes, however, Whitehead uses "fact" to refer merely to events, e.g., "The facts of life are the events of life" (PNK, p. 63).

consist simply in the repetition or persistence of various kinds of objects. Among the kinds of objects most important for natural science are: sense-objects (a particular shade of red); perceptual objects, which may be delusive (a chair perceived in a mirror) or veridical (a directly perceived chair), the latter being identical with physical objects; and scientific objects (molecules and electrons).[6] Again, unlike events, objects cannot literally speaking have parts (i.e., spatio-temporal parts). Being an abstraction, an object must be exemplified in nature (or "ingress into" nature) as a single unitary whole; often, however, one refers to the "parts" of an object in a way which derives its meaningfulness from the extended character of the events in which the object is (or may be) exemplified.[7]

The ways in which we perceive events and objects are quite distinct: "Events are lived through, they extend around us. . . . Objects enter into experience by way of the intellectuality of recognition" (PNK, pp. 63–64).[8] The following terminological distinction corresponds to the two ways of perceiving: events are said to be *apprehended*, and objects are said to be *recognized*. Events and objects are not, however, always clearly distinguished, especially by common sense. Thus, the Great Pyramid may be apprehended as one highly extended *event* (here referring to the full concreteness and inexhaustiveness—"the life of nature," as Whitehead calls it—of fact in a certain circumscribed region of space and time) or it may be recognized as a complex set of interrelated *objects* (here referring to the abstract properties which characterize or define the monument in question). An important source of the confusion between events and objects is the fact that any event becomes—and must become—an object as soon as we think or talk about it; hence it often happens that "Events are named after the prominent objects situated in them, and thus both in language and in thought the event sinks behind the object, and becomes the mere play of its relations" (CN, p. 135). What Whitehead is constantly trying to do is to rescue the events "behind" the objects from their relatively subordinate and neglected position (cf. CN, p. 77).

Two central features of Whitehead's account of sense-perception must now be briefly explained. The first is the view that the ultimate datum for sense-perception is a "specious present," that is, an experience with a finite temporal duration rather than the instantaneous experience of a moment. Like William James, whom he sometimes quotes in support on this point (see PR, p. 105), Whitehead here appeals to what he calls "instinctive" or "naïve" experience and opposes what he describes as "an intellectual theory of time as a moving

---

[6] Whitehead also sometimes mentions "percipient objects" (by which he means individual minds), but in Note I to the second edition of PNK he remarks that "the percipient object is shadowy in this book and is clearly outside 'nature' " (p. 202).

[7] But cf. Whitehead's later view: ". . . natural objects require space and time, so that space and time belong to their relational essence without which they cannot be themselves" (PNK, p. 202).

[8] The non-intellectual recognition of objects seems, however, to be more fundamental (see below, p. 147).

knife-edge, exhibiting a present fact without temporal extension" (CN, p. 68).[9] Thus, according to Whitehead, the datum of an apprehended event is the content of a specious present of some observer. This content includes much that tends to be overlooked in the usual accounts of perception. To return to an earlier example: my perception of the red patch involves the awareness of an event which is my bodily life, of an event which is the situation of the red patch, of an event which is the course of nature in my immediate environment (the total passage of nature in my room), and of a vaguely discriminated aggregate of other events (in regions outside my room).[10] This brings us to the second notable feature of Whitehead's account of sense-perception, namely, the doctrine of "significance."

Significance means the view that a full instance of sense-perception always involves two distinct but interrelated and inseparable types of awareness, "cognizance by adjective" and "cognizance by relatedness only."[11] *Cognizance by adjective* is the awareness of certain elements in an event "which are discriminated with their own individual peculiarities," while *cognizance by relatedness only* is the awareness of distant entities, undiscriminated with respect to quality and known only as bare relata in relation to the entities of the actually discerned field of events. It must be emphasized that both types of awareness are equally direct, immediate, "given." Consider once again my perception of a red patch: my awareness of the redness is cognizance by adjective; my awareness of the rest of nature as related to the redness is cognizance by relatedness. Another way of characterizing significance is to say that the field of the *discerned* (which comprises all apprehended events and all recognized objects) is only a part of the field of the *discernible* (which comprises the discerned to-

[9] In this connection it is worth quoting William James's description of the specious present: "The unit of composition of our perception of time is a *duration*, with a bow and stern, as it were—a rearward- and a forward-looking end. It is only as parts of this *duration-block* that the relation of *succession* of one end to the other is perceived. We do not first feel one end and then feel the other after it, and from the perception of the succession infer an interval of time between, but we seem to feel the interval of time as a whole, with its two ends embedded in it. The experience is from the outset a synthetic datum, not a simple one; and to sensible perception its elements are inseparable, although attention looking back may easily decompose the experience, and distinguish its beginning from its end" (*Principles of Psychology*, I, 609–10).

[10] I have deliberately chosen for discussion a more or less typical example of sense-perception. It must be emphasized, however, that Whitehead insists on the universality of the factors described in the example. It is possible to imagine artificially contrived situations in which not all of these factors would appear to be present, but further analysis will generally disclose the missing factors. Suppose, for example, that I am looking at a patch of red in an otherwise completely darkened room. Even here I would be aware of the event which is my bodily life and of an indefinitely extended region of darkness surrounding the red patch.

[11] There are also two contrasting modes of perception in the much more elaborate account of perception found in S and PR. But, though in some respects the two modes of perception distinguished in these later works ("causal efficacy" and "presentational immediacy") resemble the earlier modes, there are also many important differences between the two pairs of modes. For a brief discussion of Whitehead's later account of perception, see below, pp. 140–41.

gether with all those entities known only as necessarily related to the discerned entities). (The entire field of the discernible, relative to some apprehended event, is what Whitehead calls a "duration.")

Let us examine some further instances of significance. Suppose I am looking at an opaque sphere whose shape, color, texture, etc., I perceive in the mode of cognizance by adjective. Now, according to Whitehead, I am also *directly aware* of the existence of a geometrical center to the sphere, even though I cannot say what types of events are present at that center (e.g., the sphere may be solid or hollow). "Such knowledge," Whitehead comments, "is essentially the product of significance, since the general character of the external discriminated events has informed us that there are events within the sphere and has also informed us of their geometrical structure" (CN, p. 187). Another instance of significance is the disclosure of the space behind a mirror by means of the visual images (or adjectives) seen in the mirror. A third instance of significance is the way in which sense-objects signify perceptual objects, which are thereby known by relatedness (see UC, p. 109).

But how is cognizance by relatedness possible? Is it not grossly paradoxical to assert a knowledge of what is totally unknown—such as the inside of an opaque sphere or the space behind a mirror or the perceptual object (regarded as different from a mere class of sense-objects)? Whitehead's reply to this objection is to distinguish between the *essential* and the *contingent* relationships of any factor given in perception. The essential relationships *are* directly given along with the factor itself, but such knowledge is compatible with complete ignorance of the contingent relationships of that factor. However, even knowledge of the essential relationships is possible only if we presuppose what Whitehead calls the "uniformity" of significance. Each factor of a given type must signify *in the same manner*, i.e., must be necessarily involved in a uniform structure of relationships to other factors. The perception of redness, for example, never occurs in isolation but always in *some* (imperfectly known) *context of events* related to the event exemplifying redness.[12] Any event, moreover, is uniformly significant of other events; this may be more precisely expressed by saying that space-time relationships (which are abstractions from the mutual relationships of events) must be uniform. Similarly, the uniform significance of

---

[12] Cf. PA, p. 219: "If you ask how many other items of nature enter into the relation of crimson to cloud, I think that we must answer that every other item of nature enters into it. . . . There is no such thing as crimson lone and by itself apart from nature as involving space-time, and the same is true of cloud. The crimson cloud is essentially connected with every other item of nature by the spatio-temporality of nature, and the proposition, 'the cloud is crimson' has no meaning apart from this spatio-temporality. In this way all nature is swept into the net of the relationship.

"You may put it this way, nature as a system is presupposed in the crimsonness of the cloud. But a system means systematic relations between the items of a system. Accordingly, you cannot know that nature is a system unless you know what these systematic relations are. Now we cannot know these systematic relations by any observational method involving enumeration of all the items of nature. It follows that our partial knowledge must disclose a uniform type of relationship which reigns throughout the system."

sense-objects leads to the conclusion that any given sense-object always signifies some perceptual object which "controls" the occurrence of that sense-object. Thus, Whitehead claims he has deduced the necessity of uniform space-time and the solution to Hume's problem of induction from his doctrine of significance. Each of these deductions will be examined further below.

I have now expounded the main distinctions and doctrines of Whitehead's natural philosophy that will concern us in the succeeding chapters. It is worth remarking that, with reference to some of the great polar oppositions of philosophy, *objects* are associated with abstractness, universality, possibility, and atomicity, while *events* are associated with concreteness, particularity, actuality, and continuity.

CHAPTER IV

THE CONSTANTS

OF EXTERNALITY

1. The Philosophical Import of the Constants of Externality

We have seen that the principal types of natural elements are events and objects. Whitehead's theory of events consists of an analysis of the principal characteristics of events and a systematic application of the method of extensive abstraction to events. These topics will be considered in turn in the present and the two following chapters. This will be followed (in chapter vii) by consideration of the theory of objects, which rests in part upon certain results in the theory of events.

The principal characteristics of events are called *constants of externality;* they include:[1] (1) the fact that the observed *continuum* of nature is discriminable into a potentially definite complex of events; (2) the relation of *extension* which may hold between two events (where extension is a kind of part-whole relation among events, undefined but subsequently to be more precisely specified by a set of axioms);[2] (3) the reference of any apprehended event to a definite *duration,* namely, to that complete whole of nature (or "spatially infinite" event) "temporally" coextensive with the given event and associated with the percipient event of that perception; (4) the *percipient event* involved in the apprehension of any event; (5) the relation of *cogredience* which a percipient event bears to its associated duration, i.e., the possession by a percipient event of an absolute ("spatial") position within its associated duration; (6) the association of any event with a *community of nature* so that, for example, any percipient event always occurs in relation to "the events of antecedent or concurrent nature" which are common for all percipients. Of the six constants, only the first five figure in Whitehead's philosophy of science, since "It is unnecessary for the purposes of science to consider the difficult metaphysical question of this community of nature to all" (PNK, p. 67).

---

[1] This list is not intended to be exhaustive: "A complete enumeration of these constants is not necessary for our purpose; we need only a survey of just those elements in the apprehension of externality from which the constants of time, space and material arise" (PNK, p. 72).

[2] Cf. PR, p. 441: ". . . for the philosophy of organism the primary *relationship* of physical occasions is *extensive connection.* This ultimate relationship is *sui generis,* and cannot be defined or explained. But its formal properties can be stated."

The constants of externality are the characteristics discovered by reflection upon sensory experiences which we have "externalized," i.e., conceived as parts of external nature. What Whitehead is asking us to grasp intuitively is simply the meaning of externalizing a perception—referring it away from oneself to an independent nature.[3] It is not so much that the constants of externality are "given" in experience as that they are the products of understanding what it *means* to be "given" (from without). Essentially, to be an event means to be a given, self-sufficient fact.[4] The existence of events, then, is presupposed by all the constants of externality, which in fact must be reducible to special relations among events because "The conditions which determine the nature of events can only be furnished by other events, for there is nothing else in nature" (PNK, p. 73).

That events themselves are not directly given in perception as definite, well-defined entities is clear from the following:

> This demarcation of events is the first difficulty which arises in applying rational thought to experience. In perception no event exhibits definite spatio-temporal limits. A continuity of transition in essential. The definition of an event by assignment of demarcations is an arbitrary act of thought corresponding to no perceptual experience. Thus it is a basal assumption, essential for ratiocination relating to perceptual experience, that there are definite entities which are events; though in practice our experience does not enable us to identify any such subject of thought, as discriminated from analogous subjects slightly more or slightly less [PNK, p. 74].

Thus, an event cannot be completely divorced from the continuum out of which it has been abstracted; an event always involves relations to other events—in general, to the entire continuum of events (or "ether") which is nature, and in particular, to special events within that continuum (durations, percipient events). The consequent interrelations of the constants of externality make it impossible to describe any one of them in complete isolation from the others; thus, critics who find circularity in the description of each constant are looking for strict definitions where none is possible.

But if the constants of externality are not indubitably given in sense-perception, if they only arise after a process of reflection and analysis, how then can Whitehead maintain his anti-conventionalist, anti-bifurcationist attitude

---

[3] "The 'constants of externality' are those characteristics of a perceptual experience which it possesses when we assign to it the property of being an observation of the passage of external nature. . . . A fact which possesses these characteristics, namely these constants of externality, is what we call an 'event.'

. . . we are not considering *à priori* necessities, nor are we appealing to *à priori* principles in proof. We are merely investigating the characteristics which in experience we find belonging to perceived facts when we invest them with externality. The constants of externality are the conditions for nature, and determine the ultimate concepts which are presupposed in science" (PNK, pp. 71–72).

[4] "This conception [of nature as "given"] is the thought of an event as a thing which 'happened' apart from all theory and as a fact self-sufficient for a knowledge discriminating it alone" (PNK, p. 74).

toward the entities of natural science? It would seem that the choice of a spatio-temporal framework for natural events is not forced upon us by any of our "naïve experiences." Now, of course, Whitehead would agree that we are never directly and consciously aware in perception of precisely defined spatio-temporal relations as such relations occur in theoretical physics. However, he would insist that our experiences of nature always involve *some*, more or less vague, *extensional* aspects. In other words, we may say that externalization is a universal and necessary feature of all perception (in the sense that every per-cipient event experiences *as part of its content* extensional relations to other events), but that the isolation of this particular feature for analysis reflects the particular interests of the scientist and the natural philosopher. An alternative way of analyzing perception would be that of Part III of PR—"The Theory of Prehensions"—where the act of perception (or "actual occasion," as it is there called) is studied in its *internal* character as the datum of a specious present, hence with all temporal passage and spatiality left out. In short, for Whitehead the external world is known directly (though not, of course, in detail) as a spatio-temporally ordered structure in *every* perception.

Whitehead's anti-bifurcationism might be thought to lead him to a viciously subjectivist natural philosophy whose basic categories derive from an analysis of the highly specialized and, cosmically speaking, very rare type of events in-volved in human sense-perception. Whitehead's reply, I believe, would be that our knowledge of nature is ineluctably bound up with sense-perception; that natural science is, therefore, quite properly concerned simply with the coherence of our sense-perceptions; and that we have really no meaningful alternative to the expression of natural laws in terms of the *general* properties of those entities which become known to us in sense-perception, namely, events and objects. If this be "subjectivism," then "objectivism" could only mean an appeal to unknowable things-in-themselves. However, Whitehead's position does *not* imply that the process of sense-perception itself has any necessary relevance to natural philosophy. In fact, Whitehead asserts on the basis of a rather subtle argument that the process of consciousness is irrelevant to the analysis of nature as a process (process being, as we recall, the essence of nature in Whitehead's view). The argument runs something like this (see UC, p. 102). In the specious present the process of consciousness is suspended during the very time that nature is being apprehended as a process. But by an indefinite enlargement of the specious present we can imagine an awareness of all na-ture as a process without any process at all in the mode of awareness. Hence the process of consciousness (which, of course, includes sense-perception) may be disregarded in an analysis of the intrinsic features of nature as a process.

2. *Durations and Simultaneity*

Of the five constants of externality which are important in Whitehead's analysis of natural science, percipient events and the continuum of events require no further discussion: percipient events are familiar enough in a general way and any more detailed analysis does not belong to natural philosophy (metaphysics studies the internal character of percipient events, the sciences their external reference); while the continuum of events is so vague that nothing very definite can be said about it. Since Whitehead treats extension axiomatically in developing his method of extensive abstraction (which we shall study in detail later on), the present discussion of the constants of externality may be confined to durations and cogredience—undoubtedly two of the most problematic concepts in Whitehead's philosophy of science.

We have already noted in our preliminary survey of the constants of externality that a duration is a concrete slab of nature, finite in temporal breadth but unlimited in spatial extent. Now, such a description, if taken literally, would obviously be circular, since Whitehead will later *define* space and time in terms of durations and their properties. Let us begin, then, by considering Whitehead's statement that a duration is "a certain whole of nature which is limited only by the property of being a simultaneity" (CN, p. 53). The crucial term is "simultaneity": to justify the use of durations as part of the basis for the derivation of spatio-temporal order, Whitehead must explain how the relation of simultaneity is directly given in perception. This he does by referring to the doctrine of significance which, it will be recalled, involves the "disclosure of an entity as a relatum without further specific discrimination of quality.... Thus significance is relatedness... with the emphasis on one end only of the relation" (CN, p. 51). Specifically, "the present duration . . . is primarily marked out by the significance of an interconnected display of sensa and of other associated objects immediately apparent" (PNK, p. 203). By focusing attention upon one particular mode of significance—namely, the indefinitely extended projection of the "spatial" boundaries of a single specious present— one arrives at the notion of a duration.[5] A duration comprises all the events in nature "which share the immediacy of the immediately present discerned events. These are the events whose characters together with those of the discerned events comprise all nature present for discernment. They form the complete general fact which is all nature now present as disclosed in that sense-awareness" (CN, p. 52). "Immediacy" and "simultaneity" are synonymous but both terms must be sharply distinguished from "simultaneity" in Einstein's sense. This terminological point is clarified in the following exposition of the meaning of a duration:

There are two concepts which I want to distinguish, and one I call simultaneity and the other instantaneousness. . . . Simultaneity is the property of a group of natural elements which in some sense are components of a duration. A duration can be all nature present as the immediate fact posited by sense-awareness. A duration retains

[5] The indefinitely extended projection of the "temporal" boundaries of a specious present leads to another type of infinite event, which, however, Whitehead only came to recognize after he had completed writing CN (see the note "On Significance and Infinite Events" added in proof to CN, p. 197). For further discussion of this second type of infinite event see below, p. 56.

within itself the passage of nature. There are within it antecedents and consequents which are also durations which may be the complete specious presents of quicker consciousnesses. In other words a duration retains temporal thickness. Any concept of all nature as immediately known is always a concept of some duration though it may be enlarged in its temporal thickness beyond the possible specious present of any being known to us as existing within nature. Thus simultaneity is an ultimate factor in nature, immediate for sense-awareness.

Instantaneousness is a complex logical concept of a procedure in thought by which constructed logical entities are produced for the sake of the simple expression in thought of properties of nature. Instantaneousness is the concept of all nature at an instant, where an instant is conceived as deprived of all temporal extension [CN, pp. 56–57].

This passage merits careful examination. In the first place, it must be understood that Whitehead's concept of instantaneousness (but *not* his concept of simultaneity) and Einstein's concept of simultaneity are similar in important respects although by no means identical. Both Whitehead's instantaneousness and Einstein's simultaneity involve the idea of vanishingly small temporal breadth; Whitehead's simultaneity, on the other hand, always involves the idea of some finite, non-vanishing temporal breadth. Thus Einstein's concept of simultaneity is formulated in terms of light-signals traveling between pairs of *point*-events, while Whitehead's concept of simultaneity has nothing to do with light or with any other type of physical signal and, furthermore, applies only to actually perceivable events (hence *never* to *point*-events). Corresponding to Einstein's concept of simultaneity, Whitehead defines the concept of instantaneousness. This latter concept derives its meaning, roughly speaking, from the idea of a duration with vanishingly small temporal breadth. Such a duration—or, rather, such a limit of smaller and smaller durations—is called a "moment," which thus corresponds to the concept of an instantaneous space in physics. The precise definition of moments by means of the method of extensive abstraction we shall study later on; here the essential point is that, for Whitehead, simultaneity—a directly observable aspect of nature—is primary, while instantaneousness is derived.

We are now in a position to see more clearly how Whitehead deduces the uniformity of space-time. Moments, and ultimately, therefore, durations, are the basis of temporal and spatial relationships in nature. But since durations necessarily possess the uniformity which attaches to all perceptions in the mode of pure relatedness, it follows that the geometry of space-time must be uniform.[6] Many variations on this fundamental argument appear in Whitehead's writings. There is, for example, an argument based on our use of the uniform space-time framework of nature as a criterion—a "standard of normality"—for

[6] As Whitehead puts it: ". . . the character of our knowledge of a whole duration, which is essentially derived from the significance of the part within the immediate field of discrimination, constructs it for us as a uniform whole independent, so far as its extension is concerned, of the unobserved characters of remote events. . . . This . . . leads to the assertion of the essential uniformity of the momentary spaces of the various time-systems, and thence to the uniformity of the timeless spaces of which there is one to each time-system" (CN, p. 194).

distinguishing dreams (and presumably other illusions) from realities: if this framework is to be universally applicable to each one of our experiences, we must know its general character once and for all, and this is only possible if the framework is uniform. (See UC, pp. 102–5.) Whitehead also formulates a highly abstract version of the argument, divorced from consideration of any particular mode of significance; it runs as follows:

Every entity involves that fact shall be patient of it. The patience of fact for [any factor] *A* is the converse side of the significance of *A* within fact. This involves a canalisation within fact; and this means a systematic aggregate of factors each with the uniform impress of the patience of fact for *A*. *A* can be, because they are. Each such factor individually expresses the patience of fact for *A* [R, p. 24].[7]

The analysis of space and time is what Whitehead calls "geometry," while the description of what happens (or might happen) in space and time is what he calls "physics." Thus Whitehead says: "It is inherent in my theory to maintain the old division between physics and geometry. Physics is the science of the contingent relations of nature and geometry expresses its uniform relatedness" (R, pp. v–vi).[8] Geometry is by no means an a priori discipline; only experience can tell us, for instance, that the most appropriate geometry for use in physics is four-dimensional and Euclidean. Whitehead is perfectly willing, he says, to adopt a form of non-Euclidean (but uniformly curved) geometry if such a choice could be shown to lead to simpler formulations of physical laws. Finally, it should be noted that Whitehead's early tacit identification of his own non-causally defined durations and moments with the corresponding causally defined entities of Einstein and Minkowski[9] is later characterized in PR as a merely contingent feature of our present cosmic epoch (see below, p. 143).

[7] Cf. the following formulation in Whitehead's reply to an unfavorable review of R: ". . . my adoption of a systematically uniform geometry . . . I have based . . . on three doctrines; first, that all particular entities are abstracts from fluent fact, by which I mean the immediate unanalysed datum for knowledge; secondly, that every abstract requires a correlative systematic character throughout the datum from which it is abstracted; and, thirdly, that the particulars apparent to us in nature require systematic space-time relations throughout nature" (LPR, p. 568).

[8] This sharp distinction between essential and contingent relations in nature is slightly qualified by Whitehead later on in the text of R (p. 8): "The properties of space and time express the basis of uniformity in nature which is essential for our knowledge of nature as a coherent system. The physical field expresses the unessential uniformities regulating the contingency of appearance. In a fuller consideration of experience they may exhibit themselves as essential; but if we limit ourselves to nature there is no essential reason for the particular nexus of appearance."

[9] A moment would be represented in a two-dimensional Minkowski space-time diagram (such as Figure 2) by a straight line parallel to any of the (infinitely numerous) possible spatial axes; a duration would be represented by the region between a parallel pair of such straight lines. From the Einstein-Minkowski point of view a moment corresponds to an infinite locus of mutually simultaneous world-points, and such a locus is defined causally in the sense that only causally independent world-points can be defined as simultaneous. Durations and moments, on the other hand, are specified not in causal but in perceptual terms. Not until PR does Whitehead make use of these alternative modes of definition by distinguishing durations (causally defined) from strain-loci (perceptually defined). Cf. below, p. 143.

We may now consider Whitehead's various criticisms of Einstein's light-signal definition of simultaneity. Whitehead begins with a general criticism which would apply to any account of simultaneity construed in terms of physical signals:

. . . light signals are very important elements in our lives, but still we cannot but feel that the signal-theory somewhat exaggerates their position. The very meaning of simultaneity is made to depend on them. There are blind people and dark cloudy nights, and neither blind people nor people in the dark are deficient in a sense of simultaneity. They know quite well what it means to bark both their shins at the same instant [PNK, p. 53].

Whitehead next formulates three specific criticisms of the light-signal definition of simultaneity: (1) No determination of simultaneity is in fact ever made strictly in accordance with Einstein's definition because we live in air and not *in vacuo*. (2) There seems no justification for singling out light-signals for the definition of simultaneity instead of any one of the many other known types of physical signals (e.g., sound, water waves, nerve impulses, etc.). (3) Einstein's definition of simultaneity leaves unexplained why "The same definition of simultaneity holds throughout the whole space of a consentient set in the Newtonian group" (PNK, p. 54).[10]

All the above criticisms of the light-signal definition of simultaneity will seem at best misguided and at worst absurd unless Whitehead's special concern with meaning is borne in mind. Thus the obvious reply to Whitehead's rejection of physical signals for defining simultaneity is that *some* operational procedure for the determination of simultaneity must be used by the physicist and, hence, that physical signals become indispensable when the events in question are widely separated in space. Whitehead, however, explicitly disavows an operational theory of meaning (though not an operational theory of *measurement*), so that the real point of Whitehead's criticism has been missed. That point may perhaps be rephrased as follows. Einstein's treatment of simultaneity introduces two distinct meanings for the concept which are almost totally unrelated to one another.[11] In other words, there appears to be little in common between our intuitive apprehension of the temporal relations obtaining among the various components of a single specious present and the defined relation of simultaneity for distant events. An important symptom of the radical divergence in Einstein's two meanings of simultaneity is the fact that simultaneity at a point is a matter of direct experience, while simultaneity at distant points is based on an essentially conventional stipulation as to the behavior of light-rays. Now, Whitehead insists, as we have already noted many times, that the fundamental concepts of natural science must be based upon the immediate data of

[10] A "consentient set in the Newtonian group" is practically synonymous with what Einstein calls an "inertial system."

[11] Not totally unrelated because the light-signal definition presupposes the meaning of simultaneity at a point, i.e., the ability to recognize when two events at (approximately) the same place occur simultaneously.

sense-awareness and that in the case of spatial and temporal concepts this basis must be a *necessary* feature of sense-awareness. Instead, therefore, of attempting a define simultaneity in terms of contingent characteristics of events (namely, the dispatch and reception of light-signals by these events), Whitehead proposes to derive the concept of instantaneousness from the directly observable relation of simultaneity which is exemplified in every act of sense-awareness. However—and this is a critical point—Whitehead is *not* proposing alternatives to Einstein's methods of measuring space and time and therefore, in particular, he is *not* rejecting light-signals as an operational criterion of clock-synchronicity. In fact, Whitehead could consistently accept Einstein's rigid rods and standard clocks as devices for *measuring* space and time respectively while continuing to hold that his own definitions of spatial and temporal congruence—which are *independent of measurement*—provide those fundamental meanings of space and time which are necessarily prior to any specification of actual measuring techniques.

Turning now to Whitehead's three specific criticisms of the light-signal definition of simultaneity, let us first consider the obvious replies to these criticisms and then go on to examine the deeper significance of Whitehead's views.

1. The charge that the definition is not applicable to any real physical situation because we do not live *in vacuo* seems to ignore the fact that *all* experimental procedures are approximative in the sense that experimental conditions are never known to conform exactly to the ideal formulations of theory. Whitehead's point, however, is that the adoption of the light-signal definition makes simultaneity, and hence time itself, an elusive, never quite physically realizable concept. Surely, Whitehead would say, this is not what we *mean* by the variable t, with its range of perfectly definite numerical values, as it occurs in the equations of theoretical physics: "The ideal of accuracy shows that the meaning is not derived from the measurement." Furthermore, in Einstein's general theory of relativity, the light-signal definition of simultaneity can no longer be conceived even as an approximation to some unattainable ideal; rather, Einstein begins with a set of non-operationally definable spatio-temporal coordinate-systems (to this extent a starting point not so very unlike Whitehead's own—the difference being, of course, that Einstein's coordinate-systems do not presuppose a uniform geometry of space-time, as Whitehead's do).

2. To Whitehead's query about the reason for choosing *light* from among all possible signals for the definition of simultaneity, a sufficient answer would appear to be that light constitutes the fastest known signal (probably even the fastest physically possible signal) and consequently that the choice of light-signals to define temporal order is a necessary condition for the avoidance of temporal anomalies in connection with the behavior of other types of physical signals.[12] This can be explained more precisely as follows. It is a direct conse-

---

[12] Einstein's own comment on the choice of light-signals for the definition of simultaneity is the following: "It is immaterial what kind of processes one chooses for such a definition of time. It is advantageous, however, for the theory, to choose only those processes concerning which

quence of any definition of simultaneity which (like Einstein's) relies upon signals moving with finite speed that, for any event E, there is for a given inertial system an infinite set of other events which are temporally indeterminate with respect to E (cf. the regions labeled "Indeterminate" in Figure 2). Figure 3 represents an event E at the origin of the inertial coordinate-system K with axes x and t. Suppose that one wishes to introduce a temporal order, relative to the event E, of the sequence of events occurring at a fixed point on

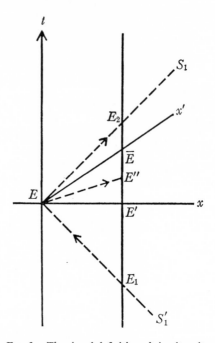

FIG. 3.—The signal definition of simultaneity

---

we know something certain. This holds for the propagation of light *in vacuo* in a higher degree than for any other process which could be considered, thanks to the investigations of Maxwell and H. A. Lorentz" (*The Meaning of Relativity*, pp. 28–29). One may wonder how what Einstein says is related to the discussion in the text above. The answer, I believe, is something like this. The electrodynamics of Maxwell and Lorentz is an extremely powerful and well confirmed theory for electromagnetic and optical phenomena occurring *in vacuo*; among other things the theory predicts that light will travel with constant speed in all directions *in vacuo*. If, now, one were to adopt some physical signal other than light (e.g., sound) for the definition of simultaneity, and if this definition forced one to conclude that light did *not* always move with constant speed—then, one would probably prefer to reject one's definition of simultaneity rather than to assume that electromagnetic theory is invalid. We have here, as also in the text below (p. 37), a case in which it is preferable to alter the definition of temporal order rather than to give up well-established scientific principles. Incidentally, the use of sound-signals to define simultaneity certainly *would* lead to some anomalies in the behavior of light, for we know that sound is *not* always propagated with constant speed even in a homogeneous medium (in particular, the motion of the medium affects the speed of sound waves moving in the medium).

the x-axis (e.g., the events $E_1, \ldots E', \ldots E'', \ldots E_2$). For this purpose, following Einstein's procedure, one would select some physical signal and *stipulate* that its speed is the same in both directions along the x-axis. Assume that possible paths of the selected signal s_1 are represented by the lines ES_1 and $S_1'E$. If we then adopt the usual (Einsteinian) definition of simultaneity, it will follow that the event E', midway between the events E_1 and E_2, is simultaneous with the event E; that E_1 and all events between E' and E_1 are earlier than E; and that E_2 and all events between E' and E_2 are later than E.[13] If we assume the existence of a physical signal s_2 whose speed exceeds that of s_1, then a possible path of s_2 will be represented by a line such as EE'' in Figure 3. Now, comparing Figure 3 with Figure 2, it is evident that any event $\bar{E}$ in the open interval E_1E_2 can be regarded as simultaneous with E in *some* inertial coordinate-system K' (assuming that simultaneity is defined in K' exactly as in K). In fact, the inertial system K' in which $\bar{E}$ and E are simultaneous will have as its x'-axis the line joining $\bar{E}$ and E, and its time-axis t' will likewise be determinate. Also, K' will obviously be moving relative to K with a speed less than that of the signal s_1.

Suppose we take as $\bar{E}$ an event *between* E'' and E_2. Then in K', E will be simultaneous with $\bar{E}$ and E'' will be earlier than $\bar{E}$, even though the signal s_2 leaves E and arrives at E'', i.e., the signal s_2 arrives at a time *prior to* the time at which it was sent. Hence, by defining simultaneity in terms of a physical signal with less than maximal velocity, we are forced to accept temporal anomalies, namely, in some cases effects will precede their causes.

I shall now attempt to develop a "Whiteheadian" rejoinder to the above defense of the choice of light (rather than some other physical signal) for defining simultaneity. Granted that physicists require *some* operational method for determining when two events are simultaneous, the question at issue becomes this: Is the time-order of events at different places merely a conventional matter? Evidence that it is not may be found in the fact that we are unwilling to accept temporal anomalies of the kind described above. In fact, the very concept of a temporal anomaly suggests that we possess an intuitive understanding of temporal order, which is, according to Whitehead, directly given in sense-perception. From the Whiteheadian point of view, the choice of light-signals for determining simultaneity represents the closest approach currently available (perhaps even the closest approach physically possible in the present cosmic

---

[13] A different stipulation concerning the relative speeds of s_1 along the two directions of the x-axis would still necessarily lead to the result that E_1 is earlier than E and E_2 later than E; however, not E' but some other event would now be simultaneous with E. In general, in a given inertial system, the time of an event E' (at point P') simultaneous with an event E (at point P) can be computed from the formula: $t' = t_1 + \epsilon(t_2 - t_1)$, where t' is the time of E', t_1 is the time of departure of an s_1-signal from P' whose arrival and reflection at P coincide with the event E, t_2 is the time of return of the reflected s_1-signal to P', and $(1 - \epsilon)/\epsilon$ is the stipulated ratio of the speeds of s_1 in the respective P'-to-P and P-to-P' directions ($0 < \epsilon < 1$).

For further details see H. Reichenbach, *Axiomatik der relativistischen Raum-Zeit-Lehre* and *Philosophie der Raum-Zeit-Lehre* (translated as *Philosophy of Space and Time*).

epoch) to the ideal implicit in our sensory knowledge of nature, namely, the ideal of an infinitely fast signal for determining simultaneity.[14] Notice that as the signal speed increases, the size of the interval E_1E_2 diminishes, so that in the limiting case of an infinitely fast signal all indeterminateness in the time-order of spatially separated events would disappear. Thus, for example, referring to Figure 3, there would be *exactly one* event, E', which could conceivably be characterized as simultaneous with E. And it is also the limiting or ideal case which guides us in selecting a value for the ratio of the signal speeds in opposite directions when we define simultaneity in accordance with Einstein's procedure; that is, we assume the ratio to be unity because only this value would, for *any* finite signal speed, lead us to describe the same unique event E' as simultaneous with E.[15]

3. Closely related to the point just discussed is the problem of how Einstein can account for the fact that his definition of simultaneity (or clock-synchronism) turns out to hold throughout a given inertial system. Einstein, in fact, simply asserts that his definition of clock-syncronism is consistent for any number of points in a given inertial system and that the relation of clock-synchronism is both symmetrical and transitive. The truth of these assertions presumably derives from certain facts concerning the propagation of light. As usual, however, Whitehead insists that even if the consistency, symmetry, and transitivity of Einstein's clock-synchronism be mere empirical facts, nevertheless our strong reluctance to accept any definition of synchronism which violates these conditions points to their origin in some fundamental, directly perceivable aspect of nature (namely, the existence of durations). Einstein's own reluctance to give up the symmetry and transitivity of synchronism is expressed in his choice of the value unity for the ratio of speeds of light in opposite directions: only for this choice are the two conditions satisfied. It is true that

[14] It should be emphasized that my discussion here is restricted to the context of a *single* inertial system; the question of the relativity of simultaneity for *different* inertial systems has not yet arisen. Whitehead believes in an *absolute* meaning for simultaneity within a single inertial system, while accepting a relativity of simultaneity for different inertial systems, as proposed by Einstein. (Cf. the discussion of Whitehead's "alternative definitions of absolute position," pp. 40–41 below.) Of course, if an infinitely fast signal were really available, then, *on the signal definition of simultaneity*, there would be an absolute simultaneity extending to all inertial systems (assuming that the speed of the signal was infinite in all inertial systems). Even in this case, however, Whitehead could still maintain a relativity of simultaneity for different observers, since for Whitehead this relativity results from the essential character of events rather than from any contingent aspects of natural phenomena (such as the existence of an upper limit to the speed of physical signals).

One may wonder how Whitehead reconciles his own refusal to base the meaning of simultaneity (and therefore of physical time) on the properties of light with his acceptance of the Lorentz-Einstein equations, which contain a universal constant whose value is the same as the speed of light *in vacuo*. Whitehead's view of this—for him—rather remarkable coincidence will be discussed in connection with his derivation of the Lorentz-Einstein equations (cf. p. 87 below).

[15] See n. 13, above, especially the formula involving ϵ. A value of unity for the ratio of the signal speeds in opposite directions obviously corresponds to $\epsilon = \frac{1}{2}$ in that formula.

Einstein gives up these conditions in his general theory of relativity, so that it is no longer possible to speak of synchronous clocks at a finite distance apart, but this is precisely one of the grounds upon which Whitehead rejects the general theory of relativity. As a matter of fact, *neither* spatial *nor* temporal relations are operationally definable in the general theory of relativity, since the very notions of a rigid rod and a standard clock are meaningless in that theory. Whitehead, to be sure, never criticizes a definition for being non-operational in character; on the contrary, from his point of view even the rigid rods of Einstein's special theory of relativity are improper primitive entities because their use—and *a fortiori* the use of the notion of "all possible quasi-rigid extensions of a rigid rod"—presupposes the genuinely primitive entities which are durations.[16]

3. Cogredience

We have noted that Whitehead derives spatial relations from durations (moments derive from durations and a moment is an instantaneous space). Temporal relations, in particular the serial order of time, are based ultimately upon the perceived succession of durations (but there is not necessarily a unique succession of durations in nature). The following passage summarizes this side of Whitehead's doctrine of time:

> We ... assume (basing ourselves upon direct observation) ... that temporal process of realisation can be analysed into a group of linear serial processes. Each of these linear series is a space-time system. In support of this assumption of definite serial processes, we appeal: (1) to the immediate presentation through the senses of an extended universe beyond ourselves and *simultaneous* with ourselves, (2) to the intellectual apprehension of a meaning to the question which asks what is now *immediately happening* in regions beyond the cognisance of our senses, (3) to the analysis of what is involved in the *endurance* of emergent objects. ... an event in realising itself displays a pattern, and this pattern requires a definite duration determined by a definite meaning of simultaneity [SMW, pp. 181–82].

The question immediately arises as to how one duration can be distinguished from another. Whitehead's answer is that "the sense of rest helps the integration of durations into a prolonged present, and the sense of motion differentiates nature into a succession of shortened durations" (CN, p. 109). But Whitehead is not content to stop here: he looks as usual for some general factor, disclosed in sense-awareness, which will ground the distinction between rest and motion. Now, it would appear to be indubitable that we can perceive our state of motion relative to the surrounding environment (although we may not know in any

---

[16] Cf. CN, p. 196: "It is to be observed that the measurement of extended nature by means of extended objects is meaningless apart from some observed fact of simultaneity inherent in nature and not merely a play of thought. Otherwise there is no meaning to the concept of one presentation of your extended measuring rod AB. Why not AB' where B' is the end B five minutes later? Measurement presupposes for its possibility nature as a simultaneity, and an observed object present then and present now."

given case whether we or our surroundings are "really" moving). For many practical purposes we single out, in a direct and intuitive fashion, that portion of our environment with respect to which we are relatively at rest. Such a three-dimensional context or background may function as a rough sort of coordinate-system with its origin at, say, the observer himself. Whatever the apparent indubitability of this view, Whitehead holds it to be an oversimplification, which, moreover, is not really able to account for the "fact" in question. The fact is that we *do* have an intuitive understanding of the meanings of motion and rest, but the distinction between them is inexplicable on the basis of an instantaneous perception of the environment. Instantaneously, rest and motion are indistinguishable; on the other hand, the content of a specious present, in which earlier and later may be compared in a single act of perception, *can* serve as the basis for the distinction of motion and rest. In Whitehead's terms, only a four-dimensional event can experience itself as "here" in its associated duration which is "now." In fact, the very possibility of durations as perceived factors in nature depends upon the ability of a percipient event to perceive directly its "preservation of unbroken quality of standpoint within the duration" (CN, p. 110). There may be change within the duration itself (which Whitehead calls "change in external nature") but this can be distinguished from change in the quality of the standpoint of the percipient event (which Whitehead calls "self-change in nature"). This special relation between a percipient event and some one duration is called *cogredience*. When cogredience holds, the percipient event will have the same temporal span as the duration and will occupy the same position in the duration throughout that temporal span.

Several aspects of the concept of cogredience call for further comment. In the first place, Whitehead holds that there is nothing peculiarly mentalistic or even biological about the concept: "Events there and events here are facts of nature, and the qualities of being 'there' and 'here' are not merely qualities of awareness as a relation between nature and mind" (CN, p. 110). In other words, not only are we aware of our own special standpoint at any given time (or better: during any sufficiently short specious present) within the present whole of nature, but we are also aware of the fact that *all* events are similarly related by cogredience to appropriate durations. For example, consider an observer in a motionless train. The train and the surrounding trees, telegraph posts, etc., are perceived as cogredient to the same duration which is the present whole of nature for that observer. But the train, trees, telegraph posts, etc., assert their own respective relations of cogredience to that duration as facts quite independent of the presence of the observer. When the train is moving, the assertion of cogredience on the part of the trees, telegraph posts, etc., is even more striking, since now these objects are *not* cogredient to the same duration as the observer himself, who therefore has a double perception of cogredience, namely, the cogredience proper to himself and the train and the cogredience proper to the trees, telegraph posts, etc.

A second thing to notice about cogredience is that it leads to a theory of absolute position, since a percipient event possesses within its associated duration a unique standpoint, one which is different from the standpoint of any other event which is a component of that duration. On the other hand, Whitehead holds that there may very well be alternative definitions of absolute position, corresponding to different meanings of simultaneity for different percipient events. We are thus led to the notion of different time-systems in nature.

. . . amid the alternative time-systems which nature offers there will be one with a duration giving the best average of cogredience for all the subordinate parts of the percipient event. This duration will be the whole of nature which is the terminus posited by sense-awareness. Thus the character of the percipient event determines the time-system immediately evident in nature. As the character of the percipient event changes with the passage of nature—or, in other words, as the percipient mind in its passage correlates itself with the passage of the percipient event into another percipient event—the time-system correlated with the percipience of that mind may change [CN, p. 111].

The possibility of alternative time-systems in nature was not seriously considered by scientists until Einstein's special theory of relativity. But the point of view from which Whitehead adopts this multiplicity of time-systems is quite different from that of Einstein. Whitehead's basic assumption is that "time is a stratification of nature." Now, although the temporal stratification of nature in each individual experience may be unique, there is good reason to believe that this uniqueness does not extend to different experiences generally. In other words, the meaning of simultaneity may be different for different individual experiences, so that two events simultaneous for one mode of stratification may not be simultaneous for another mode. Moreover, as we shall see, Whitehead contends that though scientists initially became aware of the possibility of alternative time-systems as a result of certain delicate physical experiments (the ether-drift experiments), the *special* characteristics of the phenomena involved (such as the speed of light) should not occur essentially in the formulations of an adequate natural philosophy. Rather, he argues, the mere *possibility* of many space-time systems in nature is given to sense-perception in the form of events. Events, conceived as four-dimensional volumes in which space and time are initially undiscriminated, are potentially capable of many different divisions into spatial and temporal dimensions.[17]

[17] The multiplicity of time-series also has important cosmological implications according to Whitehead: "I shall call an element of nature 'completely concrete' where, existing as it does exist, it could be all nature. . . . on the theory of the multiplicity of time-series a duration is not completely concrete; the creative advance of nature is fatally excluded if we assign two moments of one time-series as a beginning and an end, since this advance requires the whole group of time-series for its expression. . . . Thus, on the newer theory, a beginning or an end of nature within time is excluded. I regard this conclusion as a merit in the new theory" (TSM, pp. 49–50).

CHAPTER V

THE METHOD

OF EXTENSIVE

ABSTRACTION

1. Introduction

Before expounding the formal details of the method of extensive abstraction, I believe that some preliminary informal explanations may be helpful. The necessity of a systematic procedure for deriving the fundamental "abstract" entities of natural science (such as moments, points, and material objects) from factors given in sense-awareness follows immediately from Whitehead's refusal to rest content with the conception of these entities as mere "convenient fictions." Such entities are indeed convenient and are even, in a certain sense, "fictions"; but it is incumbent upon the philosopher of science to show *how* these "fictions" are related to natural entities and *why* they are convenient. Let us consider, by way of illustration, the concept of points in a plane (to simplify the discussion we ignore the fact that planes and lines are themselves ideal entities or fictions). Why is it convenient to introduce the concept of points? Whitehead's answer is something like this. Suppose that we want to express the distance between two regions of a plane surface. Clearly, the question has no unique answer; in general, the distance between the regions will vary depending upon the particular parts of each region we select as the termini of the straight line which is to serve as a measure of the distance. It is equally clear, however, that as we consider smaller and smaller regions the deviations in the various values for the distance between a pair of these regions become less marked and eventually, perhaps, negligible. If, then, we could define some sort of a minimum (perhaps "infinitesimal"?) region in the plane, we might hope that the distance between any two such regions would be uniquely determined. The usefulness (or, at least, part of the usefulness) of points is now obvious. But how shall we *define* points? The trouble with our formulation so far is that we have referred to "minimum" or "infinitesimal" regions which we have no reason to believe exist. Sense-perception always without exception reveals *finite* regions of space. Why not, then, make a virtue of necessity and define a point, not as a special sort of minimum spatial region, but as a *class* of finite spatial regions? In particular, we might attempt to define a point *P* as that class of regions which would normally be said to "converge" to *P*. An immediate objection is that there are *many* (even an infinite number) of classes of regions which "converge" to *P* (see Figure 4). Which of these classes of convergent regions—the circles, the squares, etc.—shall we use to define the point *P*? Whitehead answers: *all* of them. In effect, therefore, Whitehead proposes to define the point *P* as the complete class of all the classes of regions which would normally be said to "converge" to *P*. (Alternatively, the point *P* may be defined as the class of all regions which would normally be said to "contain" the point *P*. The two definitions are essentially equivalent.) Of course,

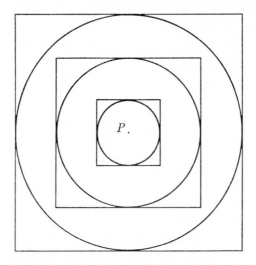

FIG. 4.—Two *K*-equal abstractive classes, each "converging" to the "point" *P*

in the formal definition of a point any reference to the point *P* must be omitted, since otherwise the definition would be circular. In fact, as already noted, Whitehead denies the existence of points construed as entities which are real in the same sense in which the (finite) members of a class of convergent regions are real; hence, as far as we know, there is literally *nothing* to which such convergent classes of regions could converge. Instead, a point is to be identified *completely* with a class of regions (appropriately chosen). It must be emphasized that our diagram (Figure 4) is misleading if it is taken to imply the existence of points as entities independent of the classes of convergent—but always *finite*—regions (the circles, the squares, etc.).

Thus, "The origin of points is the effort to take full advantage of the principle of convergence to simplicity [with diminution of extent]" (ASI, p. 206). This principle is "The master-key by which we confine our attention to such parts as possess mutual relations sufficiently simple for our intellects to consider . . .

[and it] extends throughout the whole field of sense-presentation" (ASI, p. 191).[1]

From the preceding discussion we can see that the definition of a geometrical entity such as a point requires the introduction of at least two primitive concepts, namely, the concept of a *finite region of space* and the concept of a *part-whole relation*, whose field is (finite) spatial regions and which can be used to define the notion of "converging" or "containing" regions. However, our earlier discussion was over-abstract in at least two important and related respects: first, we ignored time altogether and, second, we assumed that finite regions of space were given. To remedy these deficiencies in concreteness, we must replace spatial regions by (four-dimensional) *events;* the part-whole relation for spatial regions will then have to be replaced by the relation of *extending over* (or *extension*) for events. Here the roles of two of Whitehead's constants of externality (events and extension) have begun to emerge.

2. Extension; Junction

Whitehead begins his systematic development of the method of extensive abstraction with axioms for extension, which is a kind of whole-part relation whose field is events. On the basis of events, extension, and one other primitive concept (cogredience, also a binary relation between events), Whitehead succeeds in defining the fundamental concepts of geometry and kinematics. Two different lists of axioms for extension are formulated by Whitehead: all but one of the axioms in the later list (CN, p. 76) are included in the earlier list (PNK, pp. 101, 103); all but two of the axioms in the earlier list are included in the later list. However, Whitehead's various informal remarks concerning the properties of extension make it certain that, with the exception of one minor correction in the later list, the two lists are meant to be essentially equivalent. The following eight axioms for extension represent the conflation of Whitehead's two lists. Extension will be symbolized by K and events by lower case letters, $a, b, c. \ldots$[2]

i) If aKb, then $a \neq b$ (in other words, "part" will always mean "proper part").

[1] A somewhat more formal statement of the principle of convergence is the following: "If A and B are two events, and A' is part of A and B' is part of B, then in many respects the relations between the parts A' and B' will be simpler than the relations between A and B" (CN, p. 79).

[2] These axioms do not necessarily exhaust the properties of K; they merely formulate those properties of K essential for the definitions and theorems which are to follow. It will be noticed that Axiom v appears to introduce another primitive concept, namely, the "finitude" of an event. Since *infinite* events will be needed as an additional primitive concept later on, we might here assume that finite events are defined as non-infinite events.

For the most part, I avoid discussion of such logical properties of the axioms as independence and completeness because, as Whitehead says, "we are not thinking of logical definition so much as the formulation of the results of direct observation" (CN, p. 76).

ii) Every event extends over other events and is itself extended over by other events (the set of events extended over by an event e is called the set of "parts of e").

iii) K is transitive.

iv) If aKc, there are events such as b where aKb and bKc.

v) If a and b are any two finite events, there are events such as e where eKa and eKb.

vi) If the parts of b are also parts of a, and $a \neq b$, then aKb.

To formulate the two remaining axioms, some definitions are needed. First, however, we shall note some immediate consequences of the first six axioms. It is of some interest that the second clause in Axiom ii is redundant because in his later formulation of extensive abstraction—axiomatically very close to the earlier formulation—Whitehead omits this assumption (see below, p. 109). In fact, since any event e can always be subdivided into two events which are its "sum," we can apply Axiom v to any single event e and deduce that there are other events which extend over e—but this is simply the second clause of Axiom ii. Since K is irreflexive (i) and transitive (iii), it follows that K is asymmetrical. The field of K is compact (iv), i.e., if aKc, then there is always some event "intermediate" between a and c, "smaller than" a and "larger than" c; there are no "minimum" or "maximum" events (ii);[3] two events are identical just when their respective sets of parts coincide (by vi); and all events belong to a single manifold (v). Axioms ii, iv, and v together "postulate something like the existence of an ether [i.e., a continuum of events]" (PNK, p. 102). This continuity of nature can be more precisely characterized by the relations of "junction" and "injunction" between a pair of events, which will now be defined.

First, we define *intersection* and *separation* of two events by, respectively, the existence and non-existence of parts common to the two events. Whitehead gives two alternative definitions of junction: one of them appears first in PNK (p. 102) and is then repeated with inessential changes in CN (p. 76); the other is given in CN (p. 76). Whitehead appears to believe that the two definitions are logically independent of each other, since he holds that "If either of these alternative definitions is adopted as the definition of junction, the other definition appears as an axiom respecting the character of junction as we know it in nature" (CN, p. 76). Also, in referring to the "continuity inherent in the observed unity of an event," he asserts that "these two definitions of junction are

---

[3] It is not until his later works, beginning with SMW, that Whitehead introduces his "atomic" or "epochal" theory of time, according to which there are definite minimally extended events in nature. As a result of this innovation in his theory of time, Whitehead no longer speaks of *events* as constituting the relata of the relation of extension; instead, these relata become (in PR) "regions" in the "extensive continuum" (see p. 112). On the other hand, in one of his earliest formulations of the method of extensive abstraction (the 1916 essay, ASI), Whitehead refers to these same relata as "thought-objects of perception." The shifts in the nature of the entities which constitute the field of extension can be nicely correlated with major shifts of emphasis in Whitehead's philosophy.

really axioms based on observation respecting the character of this continuity" (CN, pp. 76–77). In fact, I believe Whitehead is mistaken about the logical relations of the two definitions, and one of them actually entails the other (but not conversely); furthermore, the weaker of the two definitions seems, as we shall see, *too* weak, considering what Whitehead most probably intends by "junction" of events. Let us look at the two definitions.

The PNK definition: Two events x and y have *junction* when there is a third event z such that (1) z intersects both x and y, and (2) there is no part of z separated from both x and y.

The CN definition: Two events x and y have *junction* when there is a third event z such that (1) both x and y are parts of z, and (2) there is no part of z separated from both x and y.

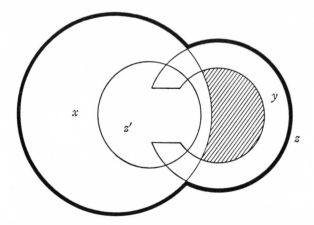

FIG. 5.—Alternative definitions of junction: x and y are joined according to the PNK definition but are not joined according to the CN definition.

The second condition is identical in both definitions and may therefore be ignored in effecting a comparison. What we find, then, is that the CN definition requires that there be an event *extending over* both the given events, while the PNK definition requires that there be an event *intersecting* both the given events. Now, the former of these two conditions is obviously stronger than the latter: an event z extending over two events x and y has by definition parts in common with x and with y, hence z must intersect both x and y; the converse is not in general true, since an event can intersect another event without extending over it. I conclude that the CN definition entails the PNK definition but not conversely. In fact, I think it can be shown that the PNK definition is too weak to define junction in the intended sense. Consider, in Figure 5, the event z (which extends over the event x and the event y) and the event z' (which intersects x and y). Since there is no part of z' which is separated from both x and y, it follows that x and y have junction in the PNK sense. On the other hand, there are parts of z (shaded in the diagram) separated from both x and y, so that *x* and *y* do not have junction in the CN sense. Assuming, then, that Whitehead would not want to say that *x* and *y* have junction in this case, I conclude that the PNK definition of junction is too weak because it does not exclude cases like those in Figure 5. That Whitehead does intend to exclude such cases seems clear from his statement that "two events with junction make up exactly one event which is in a sense their sum" (CN, p. 76)—*provided* one assumes that events with "holes" in them (e.g., *z* minus the shaded part in Figure 5) are impossible. Only much later in his development of extensive abstraction does Whitehead deal with the substance of this proviso, but he then makes it quite clear that events never have "holes" in them (see below, p. 66).

Notice that the counterexample to the PNK definition of junction (illustrated in Figure 5) depends crucially on the special shape of the event *y*. One might wonder whether events of this type (the mathematical term is "non-convex") are admissible in extensive abstraction. I believe that they are, for the two following reasons: first, such events are nowhere excluded in any of Whitehead's expositions of extensive abstraction (e.g., in PNK, CN, and PR); and second, in Whitehead's last exposition of extensive abstraction (in PR— see below p. 120), one of his principal concerns is to distinguish between convex and non-convex regions (which replace events as primitive entities in PR).

I shall assume that the CN definition of junction is adopted. Junction of events includes two distinct cases depending upon whether or not the two events *x* and *y* intersect or are separated; in the latter case *x* and *y* are said to have *adjunction*. (See Figure 6, i and ii.) Finally, the events *x* and *y* have *injunction* when (1) *xKy* and (2) there is some third event *z* separated from *x* and adjoined to *y*. (See Figure 6, iii.) Junction and injunction are obviously incompatible relations between events (to see this one must remember that "part" means "proper part"). The importance of the concepts just defined is that "Injunction and adjunction are the closest types of boundary union possible respectively for an event with its part and for a pair of separated events" (PNK, p. 103).

We are now in a position to formulate our two remaining axioms.

vii) Every event has junction (and in particular, adjunction) with other events.

viii) Every event has injunction with other events.

3. Abstractive Classes; Primes and Antiprimes; Abstractive Elements

The central conception of the method of extensive abstraction is that of an *abstractive class of events*, which is defined by the following two conditions: (1) of any two members of the class, one extends over the other, and (2) there is no event extended over by all the events in the class. It follows from these defining conditions that an abstractive class must possess an infinite number of members. The method of extensive abstraction is primarily concerned with the "convergence-properties" of such abstractive classes. Figure 4 may be taken as an illustration of abstractive classes, with the understanding that two of the dimensions of real events are missing and that in any case time is not accurately represented by an extra dimension in a spatial diagram.

Whitehead attaches great significance, as we have seen, to the view that there is no limiting event to which the members of an abstractive class converge. In our diagrams depicting abstractive classes, the points, line-segments, etc., to

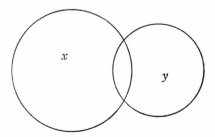

(i) Junction of intersecting events

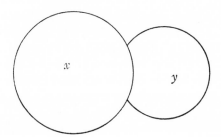

(ii) Junction of separated events (or adjunction)

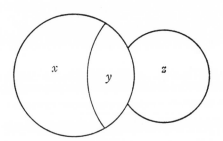

(iii) Injunction of events x and y (z is separated from x and adjoined to y).

Fig. 6.—Junction and injunction of events

which the classes of regions appear to converge do not represent real *events*, but merely various ideal entities of thought with ideally simple properties to which the members of the respective abstractive classes approximate more and more closely as one considers smaller and smaller events. Thus "the whole point of the procedure is that the quantitative expressions of these natural properties do converge to limits though the abstractive set does not converge to any limiting [event]" (CN, p. 61). The situation can be explained more precisely with the aid of a few symbols (see CN, pp. 80–81). Let e_1, e_2, e_3, $\ldots e_n$, $\ldots$ represent the successive members of an abstractive class, and let $q(e_1)$, $q(e_2)$, $q(e_3)$, $\ldots q(e_n)$, $\ldots$ represent the set of quantitative expressions which characterize the internal and external relations of the successive events. Then, says Whitehead, while the set of events s converges to *nothing*, the set of quantitative expressions $q(s)$ converges to a class of limits $l(s)$. In other words, if Q represents a particular quantitative measurement found in $q(s)$, then the homologous occurrences of Q throughout the members of $q(s)$ may converge[4] to some definite limit (one of the limits in the class $l(s)$), viz.:

$$s:\ e_1,\ e_2,\ e_3,\ \ldots e_n,\ \ldots \to \text{nothing},$$

$$Q:\ Q_1, Q_2, Q_3, \ldots Q_n, \ldots \to Q(s),$$

and in general,

$$q(s):\ q(e_1),\ q(e_2),\ q(e_3),\ \ldots q(e_n),\ \ldots \to l(s).$$

Summarizing his view of the status of abstractive sets, Whitehead says:

Thus the set s does indicate an ideal simplicity of natural relations, though this simplicity is not the character of any actual event in s. We can make an approximation to such a simplicity which, as estimated numerically, is as close as we like by

[4] Since Whitehead never claims that the law of convergence to simplicity by diminution of extent has universal and unqualified validity throughout nature (nor, therefore, that the measurable quantities Q *all* necessarily converge to limits), it would be a mistake to criticize Whitehead's method of extensive abstraction on the ground that it ignores the results of modern quantum theory. Whitehead not only admits but even stresses the fact that all natural phenomena require definite *finite* regions of space and time in which to manifest themselves. But such phenomena (involving, for example, electrons, protons, and other elementary particles) are outside the scope of geometry and kinematics, which, as Whitehead conceives them, abstract from all particular physical phenomena. Furthermore, even modern quantum theory presupposes in its mathematical formulations a continuous and therefore infinitely divisible spatio-temporal framework.

Another related criticism of Whitehead's method of extensive abstraction is that it presupposes as ultimate entities sharply bounded events which are not in fact available to sense-perception. Whitehead himself, when he was first formulating the method of extensive abstraction, raised the question of whether these ultimate entities "possess exact boundaries prior to the elaboration of exact mathematical concepts of space" (ASI, p. 213). I believe this criticism has a certain force and, while it in no way diminishes the value of what Whitehead has accomplished in his attempt to develop geometry and kinematics on a more concrete basis, it does suggest a possible direction for further technical innovations, namely, the introduction of ultimate entities which are not assumed to possess sharply defined boundaries. This same suggestion has been made by K. Menger (see Appendix II, p. 227).

considering an event which is far enough down the series towards the small end. It will be noted that it is the infinite series, as it stretches away in unending succession towards the small end, which is of importance. The arbitrarily large event with which the series starts has no importance at all. We can arbitrarily exclude any set of events at the big end of an abstractive set without the loss of any important property to the set as thus modified [CN, pp. 81–82].

Here I interpret Whitehead as follows: An abstractive set must involve a finite number of observed events at the beginning (or "big end") of the set as well as an ordering relation (namely, K) which, so to speak, "generates" indefinitely many additional unobserved (though in principle observable) events of the set. And to say, as Whitehead does, that the ideal limit approached by an abstractive set is "in fact the ideal of a nonentity" (CN, p. 61) means merely that the ideal limit is not itself a concrete event in nature. The precise ontological status of such ideal limits (as well as that of all other mathematical entities) is a problem which can only be solved within the broader philosophical context of Whitehead's later metaphysical works. Thus, for example, in PR we find that eternal objects—which constitute one of the fundamental "categories of existence"—are subdivided into those of the subjective species and those of the objective species, the latter being identified with the "mathematical platonic forms" (PR, p. 446) which "express the theory of extension in its most general aspect" (PR, p. 448). The two species of eternal objects together participate in that "unlimited conceptual realization" called the "primordial nature of God" (PR, p. 521), which thus serves as an ontological principle of explanation that is unavailable in Whitehead's natural philosophy.

Abstractive classes will in general possess different convergence-properties; thus, while the set of squares and the set of circles in Figure 4 both converge to the same point P, the set of circles β and the set of rectangles α in Figure 7 converge respectively to the point P and to the line-segment l. In order to describe with precision the convergence-properties of abstractive classes, we introduce some definitions. A class of events α is said to *cover* a class of events β when every member of α extends over some member of β. Thus, in Figure 7 the abstractive class of events α covers the abstractive class of events β, but not conversely. Covering is non-symmetrical (i.e., neither symmetrical nor asymmetrical), transitive, and reflexive. Two abstractive classes are K-*equal* (or "equal in abstractive force") when each covers the other (see, for instance, the squares and circles in Figure 4). Obviously, the relation of K-equality is symmetrical, transitive, and reflexive. Also, the relation of covering can hold, in general, between two finite classes, or two infinite classes, or a finite and an infinite class, of events; while K-equality can only hold between infinite classes of events. However, if α covers β and α is an abstractive class, then β must be an infinite class of events (because there can be no "smallest" event in β— otherwise, some of the events at the converging end of α would not extend over any event in β). In the geometrically important cases of covering, both α and β

are abstractive classes (or the closely related "abstractive elements"). *K*-equality does not, of course, imply complete identity of properties (after all, in Figure 4 for example, one abstractive class consists of circles and the other of squares): "Two equivalent [i.e., *K*-equal] abstractive sets are equivalent in respect to their convergence. But, in so far as the two sets are diverse, there will be relationships and characteristics in respect to which those sets are not equivalent, in a more general sense of the term 'equivalence' " (PR, p. 455).

A pair of important auxiliary ideas concerned with abstractive classes can now be defined (see PNK, pp. 106–7). An abstractive class is a σ-*prime* when, for some given condition σ, (1) it satisfies the condition σ, and (2) it is covered by every other abstractive class satisfying the condition σ. An abstractive class is a

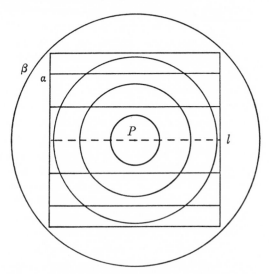

Fig. 7.—The covering of one abstractive class by another

σ-*antiprime* when, for some given condition σ, (1) it satisfies the condition σ, and (2) it covers every other abstractive class satisfying the condition σ.[5] (Any abstractive class **a** is obviously both prime and antiprime in a trivial sense with respect to the condition σ = "*K*-equal to **a**.") It follows immediately that, for a given condition σ, any two σ-primes must be *K*-equal and any two σ-antiprimes must be *K*-equal (for, according to the definitions, the two σ-primes, or two σ-antiprimes, must cover each other).

A σ-prime may be thought of as an abstractive class whose convergence is the "sharpest" consistent with the condition σ; and a σ-antiprime may be thought of as an abstractive class whose convergence is the "broadest" consistent with the condition σ. Or, as Whitehead puts it:

[5] Somewhat different definitions of primes and antiprimes are given in CN and PR. A comparison of the three sets of definitions is effected in Appendix I.

The intrinsic character of a σ-prime has a certain minimum of fullness among those abstractive sets which are subject to the condition of satisfying σ; whereas the intrinsic character of a σ-antiprime has a corresponding maximum of fullness, and includes all it can in the circumstances [CN, p. 88].

Two important questions about the logical status of σ-primes and σ-antiprimes may be raised: First, for a given condition σ, are all σ-primes (or σ-antiprimes) necessarily K-equal? Second, for a given condition σ, are there abstractive classes K-equal to a σ-prime (or σ-antiprime) but not included among the σ-primes (or σ-antiprimes)? The first question has already been answered in the affirmative; another way of expressing the result that, for a given σ, all σ-primes are K-equal and all σ-antiprimes are K-equal, is to say that every σ-prime is *both* prime and antiprime with respect to σ-primeness and that every σ-antiprime is *both* prime and antiprime with respect to σ-antiprimeness. The proof is quite trivial: since any two σ-primes (abbreviated σ_p) cover each other, it follows that every σ_p is covered by all other σ_p's, hence, every σ_p is a σ_p-prime; and analogously, every σ_p is a σ_p-antiprime. In the same way it can be shown that every σ-antiprime (abbreviated σ_a) is both σ_a-prime and σ_a-antiprime. In effect, what has been proved is that there is at most one class of K-equal σ_p's and at most one class of K-equal σ_a's.

The second question cannot be answered without further assumption (except for the trivial case when there are no σ-primes [or σ-antiprimes] at all); in other words, for a given condition σ, there may be abstractive classes K-equal to the σ-primes (or σ-antiprimes) but not themselves σ-primes (or σ-antiprimes). Thus we define a condition as *regular for primes* when (1) there are σ-primes, and (2) the set of abstractive classes K-equal to any given σ-prime is identical with the complete set of σ-primes.[6] *Regularity for antiprimes* is defined analogously. It follows immediately that if σ is regular for primes (or antiprimes), then σ_p (or σ_a) is regular for *both* primes and antiprimes. A condition σ which is regular for primes (or antiprimes) defines what would in customary logical parlance be called a "K-equivalence class" of σ-primes (or σ-antiprimes), i.e., a class of σ-primes (or σ-antiprimes) such that all σ-primes (or σ-antiprimes) within the class are K-equal to one another and no abstractive class outside the class of σ-primes (or σ-antiprimes) is K-equal to any one of the σ-primes (or σ-antiprimes).

Among the prime and antiprime abstractive classes, Whitehead attempts to single out those which are in a certain sense "absolutely" prime (or antiprime), that is, which possesses the "sharpest" (or "broadest") possible convergence. More precisely, an absolute antiprime is defined as an antiprime with respect to the condition of covering itself. It follows that an absolute antiprime is an

---

[6] This is Whitehead's definition (PNK, p. 107); in fact, a weaker definition would suffice: since it is already known that for any given σ all σ-primes are K-equal, one might define σ as regular for primes when (1) there are σ-primes, and (2) there is no abstractive class K-equal to a σ-prime, but not itself a σ-prime.

abstractive class which covers every abstractive class which covers it. The natural way of trying to define an absolute prime would be as an abstractive class which is covered by every abstractive class which it covers. Now, Whitehead points out that this latter definition is inadequate without an important qualification (see below, p. 63), but he does not seem to have noticed that his definition of absolute antiprimes suffers from a similar difficulty (see below, p. 59).

The important concept of an "abstractive element" may now be introduced. In different writings Whitehead gives two different but essentially equivalent definitions of this concept. The more direct and perhaps clearer definition identifies an *abstractive element* with the whole set of abstractive classes which are K-equal to any one of themselves (see CN, p. 84). (The important applications of this definition occur when the abstractive classes are σ-primes or σ-antiprimes [in the CN sense—see Appendix I].) The other definition (see PNK, pp. 108–9) makes use of the notions of primeness and antiprimeness, as follows. A *finite abstractive element deduced from the condition* σ is the class of events which are members of σ-primes, where σ is regular for primes (i.e., in set-theoretical terms, the union of the set of σ-primes). An *infinite abstractive element deduced from the condition* σ is the class of events which are members of σ-antiprimes, where σ is regular for antiprimes (i.e., the union of the set of σ-antiprimes). One minor disadvantage of the first definition is that abstractive elements become one logical type higher than abstractive classes (so that the relations of covering and K-equality, for example, cannot strictly apply to abstractive elements). However, a simple reformulation would seem to remedy this difficulty, viz.: define an abstractive element as *the class of events* each of which belongs to at least one of the complete set of abstractive classes which are K-equal to one of themselves; i.e., as the union of the members of the complete set of K-equal abstractive classes. (One could, of course, also reformulate the second definition in terms of *sets* of primes [or antiprimes] instead of *members* of primes [or antiprimes]; this would again have the effect of making abstractive elements one logical type higher than abstractive classes.)

The importance of abstractive elements is that they enable one to deal simultaneously with all abstractive classes possessed of a given group of convergence-properties, i.e., with complete sets of K-equal abstractive classes. This is obvious in the case of the first definition. In the case of the second definition the possible abstractive classes formed from members of a given abstractive element will include a complete set of K-equal abstractive classes (the σ-primes or σ-antiprimes) but will also in general include other abstractive classes satisfying the condition σ. Since, however, these latter abstractive classes will necessarily either cover all σ-primes [element finite] or be covered by all σ-antiprimes [element infinite], it follows that the group of convergence-properties actually determined by the abstractive element is the group possessed by the σ-primes or σ-antiprimes. Thus, according to Whitehead, an abstractive element "represents a set of equivalent routes of approximation guided by the condition that each route is to satisfy the condition σ" (PNK, p. 109).[7]

Assuming, for purposes of comparison, that the two definitions of abstractive element are so formulated that they both specify entities of the same logical type, we see that an abstractive element in the sense of the first definition is always in a trivial way both a finite and an infinite abstractive element in the sense of the second definition. For, given an abstractive class *a*, let σ be "*K*-equal to *a*." Now, σ is clearly regular for both primes and antiprimes. Also, the σ-primes are identical with the σ-antiprimes, and both are simply identical with the class of abstractive classes *K*-equal to *a*. It follows that the finite abstractive element and the infinite abstractive element (which turn out to be one and the same in this case) deduced from the above condition σ are identical with the abstractive element defined as the class of all abstractive classes *K*-equal to *a*. (Incidentally, this example shows that a single abstractive element may be *both* finite and infinite.) Finally, it should be noted that finite abstractive elements turn out to include space-time points (or "event-particles"), which are not usually thought of as finite at all. On the other hand, the most important type of infinite abstractive elements ("moments")—but by no means *all* infinite abstractive elements—are infinite in the usual sense.

On the assumption that abstractive elements are defined so as to have the same logical type as abstractive classes, an abstractive element may *cover* an abstractive class or another abstractive element. Two abstractive elements are said to *intersect* in the set of abstractive classes and abstractive elements covered by both of them. An abstractive element can evidently be *K*-equal only to itself, so that "an abstractive element has a unique abstractive force and is the construct from events which represents one definite intrinsic character which is arrived at as a limit by the use of the principle of convergence to simplicity by diminution of extent" (CN, pp. 84–85). Also, an abstractive element will be said to *inhere* in any event which is a member of it. (In CN [p. 83] inherence is defined for abstractive *classes* and events.)

4. Instantaneous Space, Space-Time, Time-less Space

Before proceeding to study in detail Whitehead's definitions for the various geometrical entities essential to the theories of natural science, I shall outline the general procedure to be followed. In the first place, Whitehead distinguishes three general types of "space": (1) the three-dimensional instantaneous (or momentary) space to which the "observed space of ordinary perception is an approximation" (PNK, p. 138); (2) the four-dimensional space-time intro-

---

[7] Note the use of the term "represents" in this passage, where the context is that of the second definition of abstractive element. In a corresponding passage within the context of the first definition of abstractive element, "represents" becomes "is": ". . . an abstractive element is the group of routes of approximation to a definite intrinsic character of ideal simplicity to be found as a limit among natural facts" (CN, p. 84). The difference in terminology is dictated by the difference in logical type of abstractive elements in the two cases.

duced by Minkowski; (3) the time-less (or permanent) three-dimensional space corresponding to the space of physical science. The second of these types of space is unique (i.e., the same for all observers); the first and third, being relative to the state of motion of the observer, are potentially infinite in number, although the third is uniquely determined for any observer in a fixed state of motion, while the first changes from instant to instant even for an observer in a fixed state of motion. Whitehead begins by defining the fundamental entities and relations in instantaneous spaces, then uses certain of these to define the fundamental entities and relations in space-time, and finally uses the latter to define the fundamental entities and relations in time-less spaces. In each case, the fundamental entities correspond to the zero-, one-, and two-dimensional geometrical objects that are usually called, respectively, "points," "straight lines," and "planes." Actually, Whitehead reserves the three foregoing terms for the fundamental entities of time-less space and he invents new terms to designate the corresponding entities in the other two types of space. The following schema should clarify Whitehead's terminology.

Instantaneous Spaces	Space-Time	Time-less Spaces
Punct	Event-particle	Point
Rect	Point-track	Straight line
	Null-track	
	Set of co-rect event-particles	
Level	Matrix	Plane
	Set of co-level event-particles	

Levels, rects, and puncts are defined in terms of "moments," which are themselves defined in terms of absolute antiprimes. Event-particles—which are fundamental to the definitions of the other space-time entities—are themselves defined in terms of absolute primes. Thus the concepts of primeness and anti-primeness are essential to the entire chain of definitions of the fundamental geometrical entities.

5. Durations; Moments; Time-systems

Earlier, in the discussion of the constants of externality, we studied Whitehead's notion of a duration as a special, undefined kind of event whose existence is assured by perception in a particular mode of significance (see above, p. 31). Whitehead also sometimes speaks of *defining* durations, e.g., "durations can . . . be defined in terms of K by [the] unlimited aspect of their extents. Namely, we assume that there are no other events with the same unlimited property" (PNK, p. 110). Now, as it stands, this is hardly a *definition* "in terms of K" of durations. However, Whitehead was perhaps thinking of defining a duration as any member of an absolute antiprime, i.e., as any member of an abstractive class which covers every abstractive class which covers it.[8] The

[8] Cf. PNK, p. 111: "Only events of a certain type can be members of an absolute antiprime, namely . . . 'durations.'"

trouble with this definition of durations is that it presupposes that there *are* such unlimited events as members of an absolute antiprime—one might just as well postulate the existence of these events from the outset by accepting durations as undefined entities. In fact, any attempt to define durations solely in terms of absolute antiprimes turns out to be impossible because, according to Whitehead's final formulation of his doctrine of significance, not *only* the events termed durations satisfy the defining characteristic of absolute antiprimes. Specifically, in the "Note: On Significance and Infinite Events" at the end of CN (pp. 197–98) Whitehead expands his doctrine of significance to include the perception, not only of durations, but also of events which are spatially finite but temporally infinite, so that there is a new set of abstractive classes—formed from the new type of infinite events—which also satisfy the definition of absolute antiprimes.[9] The relevant portion of the Note states:

. . . my limitation of infinite events to durations is untenable. . . . There is not only a significance of the discerned events embracing the whole present duration, but there is a significance of a cogredient event involving its extension through a whole time-system backwards and forwards. In other words the essential 'beyond' in nature is a definite beyond in time as well as in space. . . . This follows from my whole thesis as to the assimilation of time and space and their origin in extension [CN, p. 198].

Even more emphatic on this point is Note IV which Whitehead added to the second edition of PNK: "The attempt . . . to define a duration merely by means of its unlimitedness is a failure. In a note to the *Concept of Nature*, I point out that there is an analogous unlimitedness through time, corresponding to the spatial unlimitedness of a duration" (PNK, p. 204). Thus, we take durations as a primitive concept in accordance with Whitehead's dictum that "Sense-awareness posits durations as factors in nature" (CN, p. 59).

Parallel durations are defined as pairs of durations which are extended over by some third duration; parallelism is symmetrical, transitive, and reflexive. A *family of parallel durations* comprises all those durations parallel to a given duration including that duration itself. According to pre-Einsteinian conceptions there is just one family of parallel durations; Whitehead, on the other hand, assumes (in accordance, as he says, with the "electromagnetic theory of relativity") that there are many families of parallel durations (in other words, that some pairs of durations are extended over by other durations but that some pairs are not).[10]

[9] It is then obviously necessary to distinguish the two non-K-equal types of absolute antiprimes before they can be used to define abstractive elements; another way of putting this is to say that the condition of being an absolute antiprime is not regular for antiprimes, so that it is impossible to deduce an abstractive element from this condition alone. This difficulty can be resolved (see below, p. 72) by appealing to the distinction, based on direct sense-perception, between durations and the other, unnamed, type of unlimited events (which constitute in effect an additional constant of externality).

[10] C. D. Broad points out ("Critical Notices: *The Principles of Natural Knowledge* by A. N. Whitehead," p. 222) that one of the six axioms for extension in PNK appears to be too strong: the sixth axiom, in stating that for any two events there is a third event which extends over

We wish now to define "moments" in the sense of the following description:

A moment is to be conceived as an abstract of all nature at an instant. . . . . A moment is a route of approximation to all nature which has lost its (essential) temporal extension; thus it is nature under the aspect of a three-dimensional instantaneous space. This is the ideal to which we endeavour to approximate in our exact observations [PNK, p. 112].

In PNK (p. 110) Whitehead defines moments in terms of absolute antiprimes; for reasons given above this procedure is no longer tenable once one is prepared to admit the existence of unlimited events which are not durations. In any case, even before Whitehead accepted the existence of at least two distinct kinds of infinite events, he had already replaced his PNK definition of moments by a definition in terms of durations, as follows (see CN, p. 88). Let $\sigma =$ "being a class whose members are all durations." It follows that a σ-antiprime (in the CN sense—see Appendix I) is an abstractive class of durations which covers every abstractive class of durations which covers it. A *moment*, then, is defined as a complete set of abstractive classes each of which is K-equal to some particular σ-antiprime. It is easily seen that the members of a moment are all σ-antiprimes. Let κ be a σ-antiprime (with the above meaning for σ) and let a be an abstractive class K-equal to κ. It is clear that a must be an abstractive class of durations because only such a class could be K-equal to κ. Now, consider any other abstractive class of durations β which covers a. Since a covers κ, β must cover κ; and (by the definition of a σ-antiprime) κ must cover β. But, since a covers κ, a must also cover β. Hence, a covers every abstractive class of durations which covers it, and a is a σ-antiprime.

An important consequence of this definition of a moment is that it excludes from membership in moments abstractive classes of durations with a common "boundary" (the initial or the final boundary of all the durations). ("Boundary" will be defined below, p. 64.) This may be illustrated by a diagram. In Figure 8 the common boundary to which the abstractive class of durations β "converges" would seem to be identical with what is meant by the moment M. However, there are other abstractive classes of durations *without* a common boundary but "converging" to the same moment $M;$ e.g., a. But a and β are not K-equal, so that our definition of a moment as, in effect, a complete set of K-equal abstractive classes of durations seems inconsistent. The difficulty, however, is only apparent, since abstractive classes such as β are *not* σ-antiprimes and hence are excluded from the set of σ-antiprimes used in defining moments (a σ-antiprime [in the CN sense] must cover every abstractive class of durations which covers it: β is covered by a but does not cover a). In thereby excluding such abstractive classes as β, we "exclude special cases which are apt

---

both of them, excludes the possibility of "non-parallel" durations. This criticism seems valid, and in a later formulation of the axiom in question Whitehead does indeed restrict its application to *finite* events: ". . . given any two finite events there are events each of which contains both of them as parts" (CN, p. 76). I have included this later axiom in my list above (p. 45).

to confuse general reasoning" (CN, p. 89). Whitehead is here clearly assuming that nothing essential is being excluded; in other words, that for every abstractive class of durations like β there are other abstractive classes of durations "converging" to the same moment but lacking a common boundary for all their member-durations.

The CN definition of a moment which we have just discussed may be reformulated in terms of the PNK concept of an abstractive element. Let $\sigma =$ "covering the abstractive class of durations γ." A σ-antiprime (in the PNK sense) is, then, an abstractive class which covers γ and also covers every other abstractive class covering γ. The condition σ must now be proven regular for antiprimes, i.e., we must show (1) that σ-antiprimes exist, and (2) that the set of abstractive classes K-equal to γ is identical with the complete set of σ-antiprimes. Now, (1) cannot be proved; we must simply take it for granted that there are

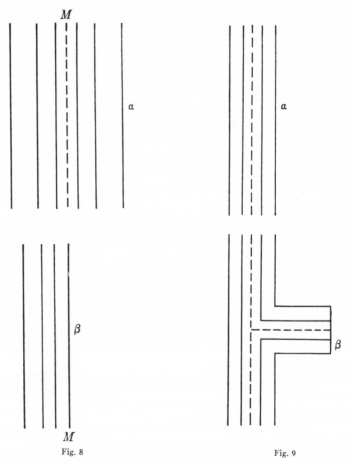

FIG. 8.—Two abstractive classes of durations, each "converging" to the moment M
FIG. 9.—Two infinite abstractive classes: β covers α but α does not cover β

such abstractive classes of durations as α in Figure 8 (notice, however, that the γ which occurs in the condition σ can be *either* of the type of α *or* of the type of β in Figure 8). But the assumption that σ-antiprimes exist was, of course, also made implicitly in the CN definition of a moment. As for the proof of (2), consider an abstractive class δ *K*-equal to a given σ-antiprime, say κ. Then δ is an abstractive class of durations covering κ; and, since κ covers γ, δ also covers γ. Thus δ satisfies the condition σ. Next, let λ be any abstractive class of durations which covers γ. Since (by the definition of a σ-antiprime) κ covers λ, it follows that δ covers λ. Hence, δ covers every abstractive class of durations which covers γ and also satisfies σ, so that δ is a σ-antiprime. Thus, σ is regular for antiprimes, and a moment is definable as the (infinite) abstractive element deduced from the condition σ.

The σ-antiprimes just used to define moments may seem to be identical with one species of absolute antiprime (which, it will be recalled, was defined as an abstractive class which covers every abstractive class which covers it). However, it is doubtful whether absolute antiprimes, as defined, can exist at all. For, consider (in Figure 9) the abstractive classes of infinite events, α and β. Clearly, β covers α but α does not cover β; thus α does not satisfy the defining condition for an absolute antiprime although it may be taken to represent an abstractive class of durations—Whitehead's own example of an absolute antiprime. The trouble seems to be that Whitehead never envisages abstractive classes like β, but it is difficult to see on precisely what basis they can be excluded.[11] Whitehead certainly seems to admit abstractive classes of *finite* events similar to β. Unless we assume that in the case of an abstractive class of infinite events we can admit only events which are durations (or some other type of infinite event with "flat" boundaries), the notion of an absolute antiprime would seem to be untenable. We have already found, however, that moments can be defined quite independently of absolute antiprimes.

Assuming that absolute antiprimes *do* exist, we note that they do not include abstractive classes of durations with a common boundary. Precisely this welcome feature of absolute antiprimes is lacking in absolute primes—which, as we shall see, complicates somewhat the definition of the abstractive elements known as "event-particles."

It can be seen that each family of parallel durations is uniquely associated with a family of parallel moments; one such family of parallel durations, together with its associated family of parallel moments, defines a *time-system*. Two different time-systems are depicted in Figure 10 (in which, as usual, two spatial

[11] Not all abstractly possible events are asserted to exist by Whitehead; for example, he does not recognize the existence of at least one type of infinite event: "It is tempting, on the mathematical analogy of four-dimensional space, to assert the existence of unlimited events which may be called the complete intersections of pairs of non-parallel durations. It is dangerous however blindly to follow spatial analogies; and I can find no evidence for such unlimited events . . ." (PNK, p. 117). Presumably the "evidence" to which Whitehead here refers consists of the deliverances of sense-perception; perhaps the same evidence is supposed to exclude infinite events like those which belong to β in Figure 9.

dimensions are omitted). The moments of a time-system are arranged serially in accordance with the following axioms:

(i) A duration belonging to a time-system is 'bounded' by a moment of the same time-system when each duration in which that moment inheres intersects the given duration and also intersects events separated from the given duration:

(ii) Every duration has two such bounding moments, and every pair of parallel moments bound one duration of that time-system:

(iii) A moment B of a time-system 'lies between' two moments A and C of the same time-system when B inheres in the duration which A and C bound:

(iv) This relation of 'lying between' has the following properties which generate continuous serial order in each time-system, namely,

(α) Of any three moments of the same time-system, one of them lies between the other two:

(β) If the moment B lies between the moments A and C, and the moment C lies between the moments B and D, then B lies between A and D:

(γ) There are not four moments in the same time-system such that one of them lies between each pair of the remaining three:

(δ) The serial-order among moments of the same time-system has the Cantor-Dedekind type of continuity [PNK, pp. 114–15].

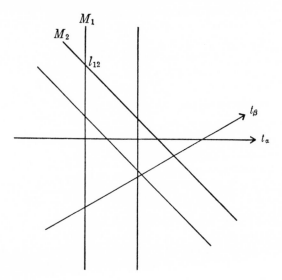

Fig. 10.—Intersection in a level of moments from two time-systems

The evidence for these axioms is "our sense-awareness of the passage of nature" (CN, p. 63); our knowledge of the direction of time derives from the same source; and the three-dimensionality of space is said to be "not an *à priori* truth but an empirical fact of nature" (CN, p. 91). The particular concept of time which he formulates is called "parabolic" by Whitehead because it is assumed to lead (as will appear below) to a "parabolic" (or Euclidean) concept of space. Whitehead suggests the possible usefulness to natural science of alternative types of time and space; in this connection he sometimes mentions both hyperbolic and elliptic space and time (e.g., R, p. v), sometimes only hyperbolic space and time (e.g., CN, p. 95). The occasional omission of the elliptic possibility represents either Whitehead's preference for an *infinite* space and time or else a mere oversight on his part. (Whitehead admits to just such an oversight involving elliptic geometry in the "Corrigenda" to PR, p. 546.)

The assumption of Cantor-Dedekind continuity for the moments of any time-system requires some comment. How, one may well ask, can the method of extensive abstraction be used to define a *non-denumerably infinite* class of entities (the moments of a time-system) on the basis of an at most *denumerably infinite* class of primitive entities (a family of durations)? Whitehead's mention of Cantor and Dedekind in this connection provides the clue to an answer: admitting the impossibility, even in principle, of detecting through sense-perception a non-denumerably infinite number of events, it follows that the moments of a given time-system must be defined as subsets of, or "cuts" in, that denumerably infinite class of moments which is, in principle at least, available to sense-perception. This is just the first of many occasions during our discussion of the method of extensive abstraction when we shall be forced to recognize the insufficiency of any interpretation of Whitehead's intent which demands that he literally deduce all geometrical entities from sense-data. Whitehead's method is, on the contrary, *constructive*, but not in an arbitrary or merely conventionalist sense.

6. Levels, Rects, and Puncts

Our next task is to define the (Euclidean) geometrical entities in each of the moments (or instantaneous spaces) of a given time-system. This is done by considering the intersections of moments from different time-systems. The locus of intersection of two non-parallel moments M_1 and M_2 consists of a set of abstractive elements and abstractive classes, and is defined as the *level* (or "instantaneous plane") l_{12} in each of the moments (see Figure 10). Consider next a third moment M_3; it may either intersect both M_1 and M_2 in the level l_{12}, or intersect only one of the previous moments (say M_1) in a new level (say l_{31}), or intersect both M_1 and M_2 in distinct levels, l_{31} and l_{23}, different from l_{12}. In the first case nothing new has emerged and the three moments are said to be *co-level;* in the second case (which occurs only when M_3 is parallel to one of the original moments) l_{12} and l_{31} are distinct levels; in the third case all three levels intersect in a common locus called the *rect* (or "instantaneous straight line") r_{123}, and the three moments are said to be *co-rect*. The consideration of a fourth moment M_4 introduces the possibility of a type of intersection different from a level or a rect, called a *punct* (or "instantaneous point"). To restrict the dimensionality of space to just three, it must be assumed that any moment either covers every member of a given punct or covers none of its members. The significance of a punct is that it "represents the ideal of the maximum simplicity of absolute position in the instantaneous space of a moment in which it lies" (PNK, p. 117).

Parallelism of levels (in a given moment) is defined by the intersection of the given moment with a set of parallel moments; *parallelism of rects* (in a given moment) is defined by the intersection of parallel levels with a given level, all within the given moment. It is a simple matter to extend these definitions to the case of non-co-momental levels and non-co-momental rects (see PNK, p. 118). Furthermore, the order of puncts on a given rect is derivative from the order of moments (in some time-system) which intersect the given rect; and since the order of puncts on a rect is found to be unique (i.e., invariant for all time-systems), the orders of moments in different time-systems may be correlated by the order of puncts on rects. Thus, finally, the set of puncts, rects, and levels within any given moment can be shown to form a complete three-dimensional Euclidean geometry—exclusive, of course, of metrical properties, which have yet to be investigated. (Such a geometry, in which parallelism but not congruence occurs, is known as "affine"—see below, n. 19.)

Whitehead's derivation of levels, rects, and puncts from moments lends point to his remark that "all order in space is merely the expression of order in time" (R, p. 60). It is surely true to say that Whitehead's entire philosophical enterprise—from natural philosophy to metaphysics—is dominated by a sense of the central importance of concepts of time and flux and change. Nevertheless, Whitehead's urgency in stressing such concepts can probably be traced in large measure to rhetorical and polemical considerations. What is certain is that any conception, like Bergson's, of "spatialization" as an inevitable vice of the intellect is totally foreign to Whitehead's thinking and, as a matter of fact, is explicitly rejected by Whitehead in his metaphysical works (see, e.g., PR, p. 319). Even the derivation of spatial from temporal order sketched above must not, if we are to heed Whitehead's advice, be taken as directly implying any deep "ontological" insights concerning the relative status of space and time within nature. Whitehead's considered judgment on the relation of space and time within nature is perhaps best stated in the following passage.

The explanation of nature which I urge as an alternative ideal to . . . [the] accidental view of nature, is that nothing in nature could be what it is except as an ingredient in nature as it is. . . . There can be no time apart from space; and no space apart from time; and no space and no time apart from the passage of the events of nature. The isolation of an entity in thought, when we think of it as a bare 'it,' has no counterpart in any corresponding isolation in nature. Such isolation is merely part of the procedure of intellectual knowledge [CN, pp. 141–42].

7. Event-Particles; Solids, Areas, and Routes

The guiding principle of the method of extensive abstraction—namely, that descriptions of natural phenomena tend to become simpler as one diminishes the extent of the spatial and temporal regions under consideration—suggests that we utilize "pointlike" or "atomic" entities for the formulation of the possible relations among natural events. Whitehead's original definition of such entities involved the notion of an "absolute prime," defined as an abstractive class covered by every abstractive class which it covers. This definition turned out to be inadequate because, as thus defined, absolute primes do not exist (see CN, p. 87). The situation is illustrated in Figure 11. Consider the two abstractive classes α and β, both of which "converge" to the same "point" P. We note that α covers β (because every member of α extends over some member of β), whereas β does *not* cover α (because it is not the case that every member of β extends over some member of α—in fact, no member of β extends over *any* member of α). We have, therefore, discovered an abstractive class, α, which covers but is not covered by another abstractive class, β; and yet α is precisely

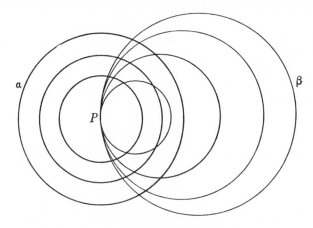

FIG. 11.—A simple abstractive class α and a non-simple abstractive class β "converging" to the same "point" P: α covers β but β does not cover α.

the type of abstractive class we would like to be an absolute prime (to be used in defining the "point" P). Hence we must attempt to define absolute primes in such a way as to exclude from consideration abstractive classes like β. This can be done by employing the previously defined puncts. We define an *absolute prime* as an abstractive class which is prime with respect to the condition σ = "covering all the abstractive classes and abstractive elements constituting some given punct." This definition eliminates β (and similar abstractive classes) from consideration in deciding whether a given abstractive class, say α, is an absolute prime because β does not satisfy the condition σ (e.g., β does not cover α). Finally, we define an *event-particle* as the (finite) abstractive element deduced from an absolute prime (the defining condition for absolute primes is clearly regular for primes).

A set of event-particles may all be covered by a single moment, in which case they are said to be *co-momental;* two event-particles which are not co-momental are said to be *sequent.* All of the event-particles on a single rect are co-momental; such a set of event-particles is arranged in an order which derives from the order of the puncts which these event-particles respectively cover. Also, any event will be said to be *analyzed* by the set of event-particles inhering in it, or to *occupy* this set of event-particles. A set of event-particles can analyze at most one event, and for every event there is a unique set of event-particles which analyze it. An event may therefore be conceived as a locus of event-particles. An event is also uniquely determined by its "boundary," which we now proceed to define.

With respect to any event *e* the manifold of event-particles may be sub-divided into three sets by considering the possible relations between *e* and the events below some definite maximum in size which belong to a given event-particle. The "sufficiently small" events belonging to the given event-particle will either all be parts of *e*, or all intersect (or "overlap") *e* without being parts of *e*, or all be entirely separate from *e*. The second case serves to characterize event-particles which "bound" *e*. More precisely: an event-particle *bounds* an event *e* when every event in which the event-particle inheres intersects both *e* and events separated from *e*. The complete set of event-particles bounding an event *e* constitutes the *boundary* of *e*. There is a one-one correlation between events and their boundaries. Events are in *contact* when their boundaries have at least one event-particle in common. Adjoined events always have partly common boundaries.

A *simple* abstractive class is defined as one for which there is no one event-particle at which all members of the converging end have contact. (The analogy between absolute primes and absolute antiprimes—provided the latter exist—would follow from the theorem that an absolute prime is a *simple* abstractive class which is covered by every *simple* abstractive class which covers it.) In most of the subsequent definitions of finite abstractive elements (e.g., solids, areas, and routes) only simple abstractive classes are admitted. This follows White-head's procedure in PNK. However, in the less formal presentation of extensive abstraction found in CN, the concept of a simple abstractive class is ignored. Since Whitehead never comments on the reason for his shift in procedure, one cannot tell whether it is due merely to the more informal context of CN or whether he really became convinced that the concept of a simple abstractive class was superfluous.

There are three aspects of an instantaneous point which must be distinguished, according to Whitehead; and furthermore, the same three aspects recur in connection with all fundamental geometrical entities:

. . . consider a point of the instantaneous space which we conceive as apparent to us in an almost instantaneous glance. This point is an event-particle. It has two aspects. In one aspect it is there, where it is. This is its position in the space. In another aspect it is got at by ignoring the circumambient space, and by concentrating attention on the smaller and smaller set of events which approximate to it. This is its extrinsic character. Thus a point has three characters, namely, its position in the whole instantaneous space, its extrinsic character, and its intrinsic character. The same is true of any other spatial element. For example an instantaneous volume in instantaneous space has three characters, namely, its position, its extrinsic character as a group of abstractive sets, and its intrinsic character which is the limit of natural properties which is indicated by any one of these abstractive sets [CN, pp. 89–90].[12]

It is possible to omit all reference to puncts once event-particles have been introduced (since a given punct defines a unique event-particle); hence Whitehead's remark that "I will always speak of 'event-particles' in preference to 'puncts,' the latter being an artificial word for which I have no great affection" (CN, p. 94). Also, levels and rects can usefully be conceived merely as loci of event-particles, namely, as the complete sets of event-particles covered by appropriately intersecting moments. This has the advantage of drastically simplifying the concepts of levels and rects, since, as originally defined, each of these types of entities will be found to include a complex variety of "linear" abstractive classes and "linear" abstractive elements (which we have yet to define).

"Solids," "areas," and "routes"—corresponding respectively to three-, two,- and one-dimensional regions in space-time—can be defined in a single uniform way by considering the intersections of pairs of spatio-temporal regions, each member of such a pair being one dimension higher than the region to be defined. However, such definitions are ill-suited to the exploration of important special types of spatio-temporal regions (e.g., "straight" routes). We shall, therefore, following Whitehead, merely sketch the general definitions of solids, areas, and routes, and then proceed to the definition of two special types of routes. Throughout the discussion, a *co-momental* (or *momental*) region will be one covered by a single moment, and a *vagrant* region will be one which is not co-momental. Also, one must bear in mind Whitehead's view that every geometrical entity possesses three distinct characters, namely, position, extrinsic

---

[12] Sometimes Whitehead refers to just two of these characters but it is not always quite clear which two he has in mind. Consider, for example, the following passage: "The peculiar simplicity of an instantaneous point has a twofold origin, one connected with position, that is to say with its character as a punct, and the other connected with its character as an event-particle. The simplicity of the punct arises from its indivisibility by a moment.

"The simplicity of an event-particle arises from the indivisibility of its intrinsic character. The intrinsic character of an event-particle is indivisible in the sense that every abstractive set covered by it exhibits the same intrinsic character. It follows that, though there are diverse abstractive elements covered by event-particles, there is no advantage to be gained by considering them since we gain no additional simplicity in the expression of natural properties.

"These two characters of simplicity enjoyed respectively by event-particles and puncts define a meaning for Euclid's phrase, 'without parts and without magnitude' " (CN, p. 94). Here "intrinsic character" seems at first to be synonymous with what is elsewhere termed "extrinsic character." A more likely interpretation, I believe, is that it is indeed extrinsic character which is being characterized in the passage *but in terms of* intrinsic character. Thus, no definite intrinsic character is ascribed to the abstractive sets covered by an event-particle; we are told merely that *whatever* intrinsic character in the way of natural properties such an abstractive set exhibits, *no* abstractive set can exhibit a simpler expression of those properties. Cf. also PNK, pp. 206–7 and CN, p. 82.

character, and intrinsic character, the method of extensive abstraction being concerned with the first two of these.

A *solid* is defined as the set of event-particles common to the boundaries of two adjoined events (see Figure 6, ii). A solid may be co-momental—in which case it is called a *volume*—or vagrant. In either case, as thus defined, a solid is a mere locus of event-particles, and as such is concerned only with illustrating a certain quality of position in the space-time manifold. To illustrate the extrinsic character of solids, they may be defined as abstractive elements. We define a *solid prime* as an abstractive class, prime with respect to the condition of being simple and covering all the event-particles common to the boundaries of the two adjoined events. This formative condition can be shown to be regular for primes, and we then define a *solid* as the abstractive element deduced from a solid prime.

An alternative definition of volumes is possible, namely, as the set of event-particles in which a moment intersects an event, provided that the set is not empty. With this meaning for volumes, *volume primes* can be defined, and then volumes as abstractive elements. These definitions are easily shown to be equivalent to the previous ones. The significance of volumes construed as abstractive elements is this:

The instantaneous volumes in instantaneous space which are the ideals of our sense-perception are volumes as abstractive elements. What we really perceive with all our efforts after exactness are small events far enough down some abstractive set belonging to the volume as an abstractive element [CN, p. 102].

Whether we ever perceive vagrant solids is for Whitehead an open question, but he suggests (without explanation) the "great importance" of such entities in Einstein's theory of gravitation (CN, p. 102). Also, Whitehead says (CN, p. 102) that the whole boundary of any finite event is a locus of event-particles constituting a vagrant solid, adding that this particular type of vagrant solid cannot be defined as an abstractive element (since no abstractive class of events can be found which will "converge" to the three-dimensional boundary of an event).[13] However, there is surely some oversight on Whitehead's part here: the whole boundary of an event could never be identical with the common boundary of two adjoined events and hence the whole boundary could never be a solid (in the sense of Whitehead's own definition). Consider, for example, an event represented by a circular disc; the complete boundary of the event in this case would be the circular circumference of the disc. But a circle could only form the common boundary of two regions if one of the regions had a "hole" in it, and we have just explained (see n. 13) how Whitehead denies the possibility of "holes"

---

[13] This example illustrates a general restriction on the types of geometrical entities definable by the method of extensive abstraction: since events are assumed to be without "holes," no entity with "holes" (e.g., a torus or the perimeter of any closed figure) can be defined by means of an abstractive class of events. In topological terms, we may say that the method of extensive abstraction is incapable of defining entities of "genus" greater than the "genus" of the primitive extensive regions (events) on which the method depends.

in events. Furthermore, the whole boundary of an event has already been defined by Whitehead and there seems no need for an additional definition.

Concerning areas, Whitehead remarks (CN, p. 103) only that *co-momental areas* may be defined as loci of event-particles by considering the overlap, if any, of the respective (two-dimensional) boundaries of a pair of volumes in a single moment, and that the abstractive elements corresponding to such areas may be defined exactly as in the case of solids. *Vagrant areas* could presumably be defined in terms of the overlapping boundaries of vagrant solids. The significance of areas construed as abstractive elements is analogous to that of volumes so construed: ". . . what we perceive as an approximation to our ideal of an area is a small event far enough down towards the small end of one of the equal abstractive sets which belongs to the area as an abstractive element" (CN, p. 103).

Routes (co-momental or vagrant) may be defined in terms of overlapping areas (co-momental or vagrant). However, it is considerably more enlightening to look upon routes as "abstractive elements in which is found the first advance [beyond event-particles] towards increasing complexity" (PNK, p. 123). Since the latter approach closely resembles that of PR (which we shall discuss in chap. vi), it merits somewhat detailed examination.

In order to introduce routes in the manner indicated, we must (as in the case of event-particles) find an appropriate condition σ, regular for primes, and then define routes as the abstractive elements deduced from the σ-primes. We begin with the notion of a *linear* abstractive class, defined as a simple abstractive class λ which (1) covers the two event-particles ρ_1 and ρ_2 (called the *endpoints*), and (2) is such that any selection from the event-particles covered by λ, which includes ρ_1 and ρ_2 but not all the event-particles, cannot be the complete set of event-particles covered by another simple abstractive class. The first condition requires that λ be more complex than an event-particle; while the second condition requires that λ possess the "linear," or one-dimensional, type of continuity (so that the exclusion of even a single event-particle from "between" the end-points leads to a locus which by virtue of its "discontinuity" could not be "converged upon" by any simple abstractive class).

Now it is obvious that, for any given pair of end-points, there will be an indefinite number of non-K-equal linear abstractive classes (corresponding to the multiplicity of possible "line-segments" between the two end-points, each "curved" in a different way). Hence, it would be clearly inadequate to take as the condition σ (for defining σ-primes) merely the general concept of linearity which has just been defined. What one wants rather is to segregate the various possible types of linearity and then to define prime abstractive classes with respect to each of these types. This is accomplished by defining a "linear prime" always *relative to* some *given* linear abstractive class. Thus, a *linear prime* is an abstractive class which is prime with respect to the twofold condition: (1) being covered by some given linear abstractive class covering two given endpoints, and (2) being itself a linear abstractive class covering the same two end-points. Recalling the definition of primeness, we note that a linear prime is an abstractive class satisfying (1) and (2) and covered by every other abstractive class satisfying (1) and (2).

We must now show that the formative condition expressed by (1) and (2) is regular for primes, i.e., we must show that linear primes exist and that any abstractive class K-equal to a linear prime is itself a linear prime. Let the given linear abstractive class referred to in (1) be λ, covering the end-points ρ_1 and ρ_2; we show first that λ itself is a linear prime, hence that linear primes certainly exist. The conditions (1) and (2) are obviously satisfied by λ (because λ covers itself and λ is by definition a linear abstractive class covering ρ_1 and ρ_2). Suppose κ to be another linear abstractive class satisfying (1) and (2); does it necessarily cover λ? First of all, κ is, by (1), covered by λ; hence the event-particles covered by κ must be a selection from the event-particles covered by λ; furthermore, by (2), the event-particles covered by κ must include ρ_1 and ρ_2. It follows that κ must cover *all* the event-particles covered by λ; otherwise κ would not be a linear abstractive class, contrary to our assumption. Thus κ does cover λ (in fact, λ and κ are K-equal), which completes the proof that λ is a linear prime.

Suppose next that μ is an abstractive class K-equal to λ; is μ necessarily a linear prime? The answer is easily seen to be in the affirmative. First of all, μ is covered by λ, so that μ satisfies (1). Also, μ covers λ and therefore μ covers ρ_1 and ρ_2; and, if λ is linear, then obviously μ must be linear, so that μ satisfies (2). Finally, we can show, exactly as for λ above, that μ must be covered by every other abstractive class satisfying (1) and (2). This completes the proof that the formative condition expressed by (1) and (2) is regular for primes.

A *route* is the finite abstractive element deduced from a linear prime. The two given event-particles which are the end-points of the linear prime from which the route is deduced are also called the *end-points* of the route, and a route *lies between* its end-points, which may be co-momental or sequent. In addition to its end-points a route will always cover an infinite number of other event-particles. The continuity of routes may be characterized by the same set of conditions already introduced to characterize the continuity of moments (see above, p. 60, Axiom iv). As Whitehead says, "The continuity of events issues in a theory of the continuity of routes" (PNK, p. 124).

Among the most important types of co-momental routes are the "straight" routes; these can be defined with the aid of rects, as follows. A *rectilinear* route is a route such that all the event-particles which it covers lie on a single rect. It is obvious that between any two event-particles on a rect there is exactly one rectilinear route. This is the definition of a rectilinear route as a locus of event-particles illustrating a certain quality of *position;* a rectilinear route can also be defined as an abstractive element. Rectilinear routes are "the segments of instantaneous straight lines which are the ideals of exact perception" (CN, p. 104).

The most important type of vagrant route is that which represents a possible path for a "material object" (this type of object will be defined below, chap. vii);

such a route is called a *kinematic* route, and it satisfies these three conditions: (1) the end-points of the route are sequent, (2) each moment, which in any time-system lies between the two moments covering the end-points of the route, covers exactly one event-particle on the route, and (3) all the event-particles on the route are covered in this way by a moment. A kinematic route may be "straight" (in the same sense in which a rectilinear route is "straight"), but no concept analogous to that of a rect is yet available by means of which such "straight paths" could be defined. At this point, then, Whitehead introduces an additional primitive concept deriving from the perceived qualities of motion and rest, namely, "cogredience."

8. Cogredience; Point-tracks and Points

We shall symbolize the binary, inhomogeneous relation of cogredience by G, so that "*eGd*" means "*e* is a finite event cogredient with the duration d." Whitehead nowhere draws up a list of axioms for cogredience, but its more important properties can be deduced from the uses to which the concept is put in the method of extensive abstraction (see PNK, pp. 128–29, and CN, pp. 111–12).

i) For every finite event e there is a unique duration d such that eGd.

ii) If eGd, there are finite events such as e' where eKe' and $e'Gd$.

iii) If eGd, there are finite events such as e' where $e'Ke$ and $e'Gd$.

In order to formulate the next axiom the following definition is introduced: A finite event e *extends throughout* a duration d when (1) dKe, and (2) e is intersected by any moment which inheres in d.

iv) If eGd, then e extends throughout d.

It follows from Axiom iv that (*a*) if eGd, then dKe; and (*b*) if eGd, then both the initial and final boundary moments of d cover some event-particles on the boundary of e. Using the consequent clauses of (*a*) and (*b*) to define: *a finite event e begins with duration d and ends with it*, we find that if eGd, then e begins with d and ends with it.

The following important theorem can now be proved:[14]

Theorem 1. If eGd and dKd', then d' intersects e in an event e'
such that eKe' and $e'Gd'$.

Consider an event-particle P which inheres in a given duration d. There will be a set of events in which P inheres and which are cogredient with d. Each of these events occupies its own aggregate of event-particles; we can call the class

---

[14] A formal proof would be straightforward but tedious, so I shall follow Whitehead in omitting it. There is (I assume) a typographical error in Whitehead's formulation of the theorem (CN, p. 112, last four lines from the bottom): the minimum change necessary to correct the grammar while preserving the punctuation in the text would be to change the first word "If" to "Let" (or some synonym).

of event-particles common to all these aggregates the *station* of P in d. This is the definition of a station as a locus. The definition of a station as an abstractive element may be constructed as follows. We first define a *stationary prime* within a duration d as a prime which is a simple abstractive class, such that each of its members extends over events which (1) are inhered in by some assigned event-particle P inherent in d, and (2) have the relation G to d. This formative condition is regular for primes. The abstractive element deduced from a stationary prime within a duration d is the *station* within d.

The foregoing definition of a stationary prime is illustrated in Figure 12. The

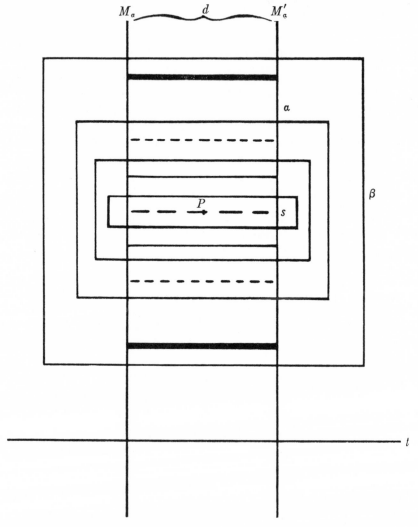

FIG. 12.—The station s of the event-particle P in the duration d bounded by moments M_a and M'_a (two dimensions of M_a and M'_a are omitted).

essence of the definition can be seen to lie in the existence of abstractive classes (such as a) which "converge" to the station s: it is the members of a which satisfy conditions (1) and (2) in the definition. However, a is not a *simple* abstractive class (because all of its members have a partly common "boundary"); hence a stationary prime must be defined in terms of abstractive classes such as β, which are prime with respect to the formative conditions of being simple and covering a. In CN, where Whitehead throughout ignores the concept of simple abstractive classes, a stationary prime is defined merely as an abstractive class which is prime with respect to the conditions (1) and (2) of the PNK definition; thus, a in Figure 12 is a stationary prime in the CN sense.

Whitehead introduces at this point (without proof) what he calls "the fundamental theorem" (PNK, p. 129):

Theorem 2. If d and d' are durations of the same time-system such that dKd', and if the event-particle P inheres in d', and s and s' are the stations of P in d and d' respectively, then s covers s' [in less technical terms: if d' is part of d, then s' is part of s].

I shall sketch a proof of this theorem using the CN definition of a station. (It is easy to see that if the theorem holds for the CN definition of a station, it must also hold for the PNK definition of a station.)

Assume that a is one of the stationary primes defining the station s of the event-particle P in the duration d; likewise, assume that a' is one of the stationary primes defining the station s' of P in d', where dKd'. We must prove that a covers a'. To do this we shall find an abstractive class $\bar{a}$ which covers a' and is covered by a.

Consider now those events which are members of a; each of them (say, e) intersects d' in an event (say, $\bar{e}$). By Theorem 1 $eK\bar{e}$, so that the class $\bar{a}$ of events such as $\bar{e}$ is covered by a. Next, we shall show that $\bar{a}$ is an abstractive class each of whose members (such as $\bar{e}$) satisfies the two conditions: (1) $\bar{e}Gd'$, and (2) P inheres in $\bar{e}$. That $\bar{a}$ is an abstractive class is evident from the fact that it consists of the intersections of the events in the abstractive class a with a single event, d' (a theorem to this effect could easily be proved on the basis of our previous definitions and axioms for extension). That $\bar{e}Gd'$ follows from Theorem 1. Finally, since P inheres in e and in d', P must inhere in the intersection of e and d', namely, $\bar{e}$.

Now, a' is an abstractive class prime with respect to the conditions (1) and (2) above; hence a' is covered by any other abstractive class satisfying those conditions, so that $\bar{a}$ covers a'. We conclude that a covers a' (a covers $\bar{a}$, $\bar{a}$ covers a', and the relation of covering is transitive).

Theorem 2 and the fact (implicit in the definition of a station) that each event-particle in a duration has a uniquely determined station in that duration, enable us to prolong indefinitely any given station in a duration d throughout the time-system to which d belongs. The complete locus of event-particles as thus defined by a station is a *point-track* (in a given time-system). Point-tracks in the same time-system never intersect and so are called *parallel*. Also, parallelism of point-tracks and parallelism of rects are interconnected, e.g., a set of parallel point-tracks which intersect a rect *r* lying in a moment *M* will intersect any moment parallel to *M* in a rect parallel to *r*. Analogous theorems hold for a point-track and a family of parallel rects, and for two families of point-tracks.

The possibility of an alternative (and considerably simpler) definition of point-tracks emerges with Whitehead's admission of temporally infinite events analogous to durations (see p. 56 above):

It follows from this admission that it is possible to define point-tracks . . . as abstractive elements. This is a great improvement as restoring the balance between moments and points. . . . This correction does not affect any of the subsequent reasoning in . . . [PNK and CN] [CN, p. 198].

A point-track could then be defined, in complete analogy with the definition of a moment, as (roughly) an infinite abstractive element, all of whose members belong to abstractive classes composed entirely of infinite events (of the new kind) belonging to a single parallel family.

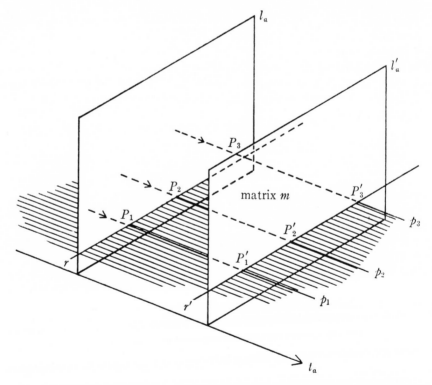

Fig. 13.—Point-tracks and points; matrices and straight lines. Each point-track p_n defines a point in time-less space of t_a; the set of parallel point-tracks p_n in the matrix m define a straight line in time-less space of t_a.

The *points* in a time-less space are defined as the point-tracks of a parallel family. This is illustrated by Figure 13, in which l_a and l'_a represent parallel "cross-sections" through the moments M_a and M'_a of the time-system a (in other words, l_a and l'_a are parallel levels in the moments M_a and M'_a respectively); P_1 and P'_1 represent event-particles in M_a and M'_a respectively, such that $P_1P'_1$ ($=p_1$) is a point-track associated with the time-system a; and therefore p_1 represents a point in the time-less space of a. Similarly, $P_2P'_2$ ($=p_2$) and $P_3P'_3$ ($=p_3$) represent two more points in the time-less space of a.

Let us recapitulate our procedure in defining a point in time-less space: what we have done in effect is to use the relation of cogredience to select a set of event-particles, one from each moment of a single selected time-system. The intuitive meaning of this set of event-particles (which constitutes the point) is that anything (such as a "material particle") which occupies the members of such a set is at rest within the given time-system. Hence, though motion and rest are relative to the time-system, there is something like absolute position *within* each time-system; in Whitehead's words:

. . . you cannot have a theory of rest without in some sense admitting a theory of absolute position. It is usually assumed that relative space implies that there is no absolute position. This is, according to my creed, a mistake. The assumption arises from the failure to make another distinction; namely, that there may be alternative definitions of absolute position. This possibility enters with the admission of alternative time-systems. Thus the series of spaces in the parallel moments of one temporal series may have their own definition of absolute position correlating sets of event-particles in these successive spaces, so that each set consists of event-particles, one from each space, all with the property of possessing the same absolute position in that series of spaces. Such a set of event-particles will form a point in the timeless space of that time-system. Thus a point is really an absolute position in the timeless space of a given time-system [CN, 105–6].

9. Matrices, Straight Lines, and Planes; Null-tracks

As point-tracks were needed to complete our theory of "straight lines" in space-time (since rects provide us only with co-momental "straight lines"), so new entities (called "matrices") are needed to complete our theory of "planes" (since levels provide us only with co-momental "planes"). A *matrix* can be defined in terms of rects, event-particles, and point-tracks: consider a rect r and an event-particle P non-co-momental with r; by forming the locus of event-particles on rects or point-tracks through P and intersecting r, including the event-particles on the rect through P and parallel to r, one obtains a matrix (e.g., the event-particle P'_2 and the rect r in Figure 13 define the matrix m "containing" the rects r and r' and the point-tracks p_1, p_2, and p_3). Alternatively, a matrix can be obtained from an event-particle P and a point-track p not passing through P (e.g., the event-particle P'_2 and the point-track p_1 in Figure 13 also define the matrix m). Any matrix can be generated in either of these two ways. *Parallelism of matrices* and its relationship to parallelism of levels can be developed on the basis of the foregoing definitions. Furthermore, just as a point-track constitutes a point in the time-less space of some time-system, so the set of points defined by a set of parallel point-tracks within a given matrix constitutes a *straight line* in the time-less space of some time-system (e.g., the set of points defined by the point-tracks p_1, p_2, p_3, etc., in Figure 13 define a straight line in the time-less space of t_a, namely, the straight line which is "occupied," successively, by the rects r, r', etc.).

We are now in a position to complete our theory of "straight lines" in space-time by defining a set of entities which may be thought of indifferently as rects

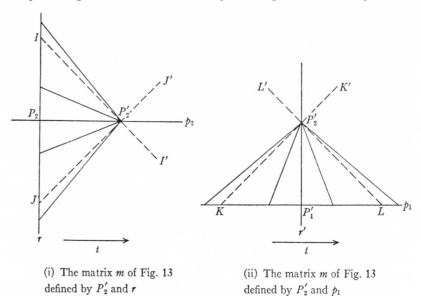

(i) The matrix m of Fig. 13 defined by P_2' and r

(ii) The matrix m of Fig. 13 defined by P_2' and p_1

FIG. 14.—The relation between rects, point-tracks, and null-tracks within a given matrix

or point-tracks. Consider the matrix m in Figure 13 defined, first, by the event-particle P_2' and the rect r, and then by the event-particle P_2' and the point-track p_1 (the two cases are illustrated respectively by Figure 14, i and ii). It is readily seen that, relative to the event-particle P_2', the rect r and the point-track p_1, respectively, are divided into three mutually exclusive segments by the pair of event-particles I, J, and the pair of event-particles K, L. Thus all event-particles between I and J can be joined to P_2 by point-tracks (e.g., p_2) and all event-particles beyond I or J can be joined to P_2' by rects; while all event-particles between K and L can be joined to P_2' by rects (e.g., r') and all event-particles beyond K or L can be joined to P_2' by point-tracks. Hence a matrix is divided, relative to an event-particle in it, into four mutually exclusive regions by two loci (II' and JJ' respectively coinciding with LL' and KK' in Figure 14, i and ii) which may equally well be called rects or point-tracks— these loci (and all others parallel to either one of them) Whitehead terms *null-*

tracks. In any matrix there are two families of parallel null-tracks, exactly one member of each family passing through each event-particle in the matrix. Thus, through any given event-particle there is an infinity of null-tracks which form two three-dimensional "conical" surfaces whose mutual vertex is the given event-particle. The order of event-particles on a null-track is derived from its intersection with systems of non-co-momental parallel rects or of parallel point-tracks, or from the order of event-particles on routes lying on the null-track.

Recalling our discussion of the space-time of special relativity (chap. ii), we see that Whitehead's rects, point-tracks, and null-tracks correspond precisely to "straight lines" in space-time which are, respectively, space-like, time-like, and null. Since we have defined parallelism for all three types of "straight lines," we have established what is termed an affine geometry for space-time (see below, n. 19).

Finally, to complete our account of time-less spaces we must define two-dimensional "flat" spatial entities. Whitehead seems to have omitted this definition from his systematic development in PNK, but the missing definition is formulated in CN (p. 118): a *plane* in the time-less space of a time-system a is the locus of all those points of the space of a which intersect any moment M of a in event-particles lying on a single level in M. Thus, in Figure 13 the set of all point-tracks in the matrix m and in all matrices parallel to m would constitute a plane in the time-less space of t_a.

We have now characterized completely the non-metrical aspects of the time-less, Euclidean, three-dimensional space of theoretical physics, and our next step is to derive the metrical aspects of this space. Whitehead is most insistent about the logical priority of non-metrical over metrical geometry:

. . . it must be remembered that measurement is essentially the comparison of operations which are performed under the same set assigned conditions. If there is no possibility of assigned conditions applicable to different circumstances, there can be no measurement. We cannot, therefore, begin to measure in space until we have determined a non-metrical geometry and have utilized it to assign the conditions of congruence agreeing with our sensible experience [ET, p. 247].

Whitehead's attempt to fulfil this program does not come, as we shall see, until his further development of the method of extensive abstraction in PR. The desideratum of Whitehead's earlier treatment of geometry (the subject of this chapter) is a derivation of the common non-metrical properties of all uniformly curved spaces (elliptic and hyperbolic, as well as Euclidean); this derivation will be examined in chapter vi.

10. Congruence

The problem of defining congruence assumes a special significance for Whitehead because, as we have already seen, he rejects those theories of meaning which would reduce congruence to measuring operations. Now, congruence, according to Whitehead, is founded on "repetition," which is itself a characteristic of parallelism and "normality": in the case of parallelism, the property of repetition is illustrated by the theorems on parallel rects and point-tracks stated above; in the case of normality, we have symmetry about the normal, and symmetry is itself a kind of repetition. Hence we might expect to discover the nature of congruence in some properties of parallelism and normality. Thus, we begin by defining normality. A point-track is defined as *normal* to the moments of the time-system in the time-less space of which it is a point. A matrix is defined as *normal* to the moments which are normal to any of the point-tracks which it contains. The mutual normality of levels and matrices can also be defined (see PNK, pp. 139, 205). Finally, if *l* and *m* are respectively a level and a matrix normal to each other, then the rects in *l* will be called normal to the rects and point-tracks in *m*. A pair of normal rects are called *perpendicular* (or *at right angles*).

We look now for various "repetition properties," associated with the parallelism and normality of straight routes, which may serve to specify a meaning for the "equality" or "congruence" of any two segments of such routes. Since a straight route may lie on a rect or on a point-track, there are several different cases to be considered.[15] We begin with routes along the same or parallel rects and point-tracks, then proceed to routes along non-parallel rects, and finally come to routes along non-parallel point-tracks. The following three axioms are concerned explicitly with congruence of routes along parallel rects or along parallel point-tracks:

i) The opposite sides of a parallelogram are congruent.
ii) Congruence in the sense of Axiom i is a symmetrical relation.
iii) Congruence in the sense of Axiom i is a transitive relation.[16]

[15] I follow here the *general* line of thought of PNK, pp. 141–46, but I have altered some details in the interests of clarity and precision. Whitehead's treatment of congruence in PNK suffers from carelessness in its formulation of axioms, and the later discussions of congruence in CN (pp. 128–30) and R (pp. 51–57) offer no marked improvements. A minor evidence of Whitehead's carelessness is his allusion to the definition of congruence given by E. B. Wilson and G. N. Lewis (see n., p. 141, PNK)—a definition which is nowhere to be found in the work cited.

[16] Cf. PNK, p. 141. Whitehead's statement that "the opposite sides of parallelograms are congruent to each other" is probably meant to be equivalent to my Axioms i and ii; while his statement that "two routes which (as thus defined) are congruent to a third route, are congruent to each other" and my Axiom iii are equivalent provided one assumes congruence to be symmetrical. Whitehead refers to his own statement of the transitivity of congruence as "a substantial theorem as to parallelism, and not a mere consequence of definitions" (PNK, p. 141)—by which, presumably, it is meant that the assumption of transitivity is not derivable from Whitehead's other assumptions (and hence is *not* a "theorem," in the usual sense of the term, which opposes "theorems" to "axioms"). Whitehead also formulates what amounts to a special (and therefore superfluous) case of my Axiom iii and of his own more general statement quoted above: ". . . routes on the same rect, or on the same point-track, which are congruent to the same route are congruent to each other" (PNK, p. 141).

It is obvious from the definitions of rect and point-track that opposite sides of any parallelogram will be either both rects or both point-tracks. Thus we have provided for the comparison of routes on parallel rects and parallel point-tracks (routes along a single rect or a single point-track constitute a special case of our general axioms—see Figure 15). Since all point-tracks associated with a given time-system are parallel, we have a complete account of the congruence of time-lapses *within* a given time-system; we have yet to formulate axioms for the congruence of time-lapses in *different* time-systems. On the other hand, our account of spatial congruence is not yet complete even for the set of rects within a single time-less space: if two rects are non-parallel (i.e., either intersecting or skewed), there is as yet no way of comparing the lengths of routes on these rects. This gap will be filled by the next set of axioms.[17]

iv) If in a co-level triangle the median and the altitude to a particular side coincide, then the other two sides are congruent.

v) Congruence in the sense of Axiom iv is a symmetrical relation.

vi) Congruence in the sense of Axiom iv is a transitive relation.

In a co-level triangle all three sides must be rects. We shall now see how the Axioms iv–vi make all rects comparable with respect to the lengths of routes along them. If the two routes in question possess an end-point (defined by an event-particle) in common, one has simply to complete the triangle by a segment of a third rect and then see if the median to this third rect is also an altitude. Suppose, on the other hand, that the two routes in question do not possess an end-point in common; for the sake of definiteness let us assume they do not intersect at all. Let *AB* and *ED* in Figure 16 represent the two straight routes whose possible congruence is being determined. First, we construct a rect parallel to *ED* and passing through *A*; then we complete the parallelogram *ACDE*. By Axiom i, *AC* and *ED* are congruent. Next, after completing the triangle *ABC*, we determine if the altitude and the median on *BC* coincide (we have already adequate criteria for the perpendicularity of *BC* and *AM* and for the equality of *BM* and *MC*). Thus we can discover whether or not *AB* and *AC* are congruent (by Axiom iv) and hence whether or not *AB* and *ED* are congruent (relying on Axioms v and vi).[18]

Suppose now that we wish to compare routes along rects which are in different but parallel moments. Again it is obvious that our six axioms suffice. A

[17] Cf. PNK, p. 142. Whitehead's statement is more general than my Axiom iv, since he admits co-matricial as well as co-level triangles. However, in his subsequent discussion he explains how only the co-level case is required: the other case admits four distinct possibilities, two of which are derivable from the co-level case and the other two of which are better treated by means of quite different axioms concerning relative motion. (For an analysis of the five possible types of triangles which closely follows Whitehead's, see my discussion in the text below.) As for my Axioms v and vi: Whitehead once more implicitly assumes the symmetry of congruence and he holds that "the transitiveness of congruence expresses a substantial law of nature and not a mere deduction from the terms of the definition" (PNK, pp. 142–43).

[18] The construction I have given is similar to that proposed by Whitehead (cf. PNK, p. 145, Figure 15).

somewhat more complicated case occurs when the two rects are in non-parallel moments. But here too there is no real difficulty: one can always find a pair of intersecting moments, respectively containing the two rects in question; and any rect in the level common to the two moments is comparable to *both* of the original rects. Thus we have obviated the need for any additional axioms respecting co-matricial triangles. The whole situation is clarified by Figure 17,

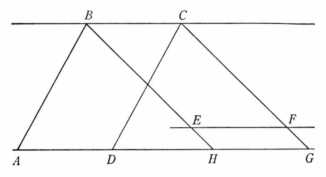

AD congruent BC,
EF congruent BC,
HG congruent BC;
therefore, AD congruent EF,
and AD congruent HG.

FIG. 15.—Congruence as derived from parallelism

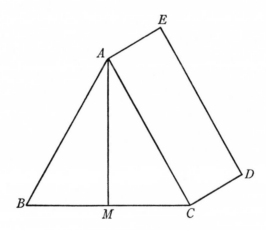

ED congruent AC,
AB congruent AC;
therefore, AB congruent ED.

FIG. 16.—Congruence as derived from normality

which represents the five possible types of triangles as determined by the nature of their respective sides (rects or point-tracks). Thus far we have accounted for cases i, ii, and iii, but not for iv and v. These two latter cases involve time-congruence between *different* time-systems; after we have analyzed motion, axioms concerning relative velocity will be introduced to secure the universal comparability of time-lapses. And finally, it should be noted that we have still to discuss the congruence of routes along null-tracks; it will be convenient to postpone this last topic until the various possible types of kinematics have been investigated (see below, p. 84).

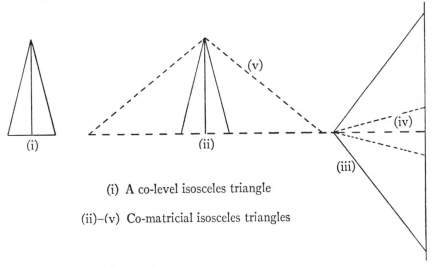

(i) A co-level isosceles triangle

(ii)–(v) Co-matricial isosceles triangles

Fig. 17.—The possible types of isosceles triangles. (Solid lines indicate rects, broken lines point-tracks.)

11. Motion; Kinematics

We now find ourselves in a position to set up Cartesian coordinate-systems and thereby to formulate a four-dimensional analytic geometry for space-time. More specifically, consider a time-system α: through any arbitrarily chosen event-particle O (called the origin) we can find a set of three mutually perpendicular α-rects (call them $OO_{\alpha x}$, $OO_{\alpha y}$, and $OO_{\alpha z}$) and an α-point (call it $OO_{\alpha t}$) normal to all three α-rects. These four axes may be used to determine the coordinates of any event-particle in the space-time of α. For any other time-system β, another set of mutually normal axes, $OO_{\beta x}$, $OO_{\beta y}$, $OO_{\beta z}$, $OO_{\beta t}$, can be found, and the two sets can obviously be adjusted so that $OO_{\alpha y}$ is identical with $OO_{\beta y}$, and $OO_{\alpha z}$ with $OO_{\beta z}$. Axes so chosen are called *mutual axes* for the two systems α and β; the coordinates of any event-particle with respect to the system α are $(x_\alpha, y_\alpha, z_\alpha, t_\alpha)$ and with respect to the system $\beta(x_\beta, y_\beta, z_\beta, t_\beta)$, where always $y_\alpha = y_\beta$ and $z_\alpha = z_\beta$. It is to be noted that lengths on all rects, either in α or β, are measurable in terms of one unit length, but time-lapses between α-moments—or along α-points—must be measured in a time-unit peculiar to the time-system α, and likewise for β.

The formulae for transformation from α-coordinates to β-coordinates, referred to mutual axes for simplicity, are of the form:

$$
\left.
\begin{aligned}
x_\beta &= \Omega_{\alpha\beta} x_\alpha + \Omega'_{\alpha\beta} l_\alpha \ , \\[4pt]
y_\beta &= y_\alpha \ , \\[4pt]
z_\beta &= z_\alpha \ , \\[4pt]
l_\beta &= \Omega''_{\alpha\beta} l_\alpha + \Omega'''_{\alpha\beta} x_\alpha \ ,
\end{aligned}
\right\} \quad (1)
$$

where the Ω's are constants dependent on the two systems and on the two arbitrarily chosen time-units in α and β, but not dependent on the arbitrarily chosen rects OO_{ay} and OO_{az}. This form for the transformation equations is justified as follows. The relation between y_α and y_β must be one-one, and y_α and y_β must vanish together; hence y_β must be a constant multiple of y_α. Since the relative velocities of α and β are equal and opposite along the coincident x_α- and x_β-axes, there is no component of velocity along the y_α- or y_β-axes; hence in the transformation equations for y_β ($y_\beta = k_{\beta\alpha} y_\alpha$) and $y_\alpha(y_\alpha = k_{\alpha\beta} y_\beta)$, the constants $k_{\beta\alpha}$ and $k_{\alpha\beta}$ must, by symmetry considerations, be equal; which implies that $y_\beta = y_\alpha$. A precisely analogous argument leads to the result that $z_\beta = z_\alpha$. Furthermore, the equations for x_β and l_β cannot involve either y_α or z_α because of the homogeneous and isotropic character of our three-dimensional Euclidean space. Also, these equations must be linear (in x_α and l_α) because of the affine character of space-time and the fact that affine (or parallelism-preserving) transformations are linear.[19] The corresponding set of equations for transformation from β-coordinates to α-coordinates is:

$$
\left.
\begin{aligned}
x_\alpha &= \Omega_{\beta\alpha} x_\beta + \Omega'_{\beta\alpha} l_\beta \ , \\[4pt]
y_\alpha &= y_\beta \ , \\[4pt]
z_\alpha &= z_\beta \ , \\[4pt]
l_\alpha &= \Omega''_{\beta\alpha} l_\beta + \Omega'''_{\beta\alpha} x_\beta \ .
\end{aligned}
\right\} \quad (2)
$$

[19] The general form of an affine transformation in three variables is:

$$
\begin{aligned}
x' &= a_1 x + b_1 y + c_1 z + d_1 , \\
y' &= a_2 x + b_2 y + c_2 z + d_2 , \\
z' &= a_3 x + b_3 y + c_3 z + d_3 ,
\end{aligned}
$$

where the a's, b's, c's and d's are constants. Physically, such a transformation corresponds to a "homogeneous deformation" (i.e., a homogeneous stretching or contraction). When the coefficients of x, y, and z satisfy certain conditions (for which see Appendix III, equation [2]), the transformations become *metrical* (or distance-preserving). For an introduction to the subject of coordinate transformations see F. Klein, *Elementary Mathematics from an Advanced Standpoint*, Vol. II, *Geometry*.

Whitehead says that these two sets of equations must be "equivalent,"[20] which implies that the following conditions must be satisfied:

$$\frac{\Omega_{\alpha\beta}}{\Omega''_{\beta\alpha}} = \frac{\Omega''_{\alpha\beta}}{\Omega_{\beta\alpha}} = -\frac{\Omega'_{\alpha\beta}}{\Omega'_{\beta\alpha}} = -\frac{\Omega'''_{\alpha\beta}}{\Omega'''_{\beta\alpha}} = \frac{1}{\Omega_{\beta\alpha}\Omega''_{\beta\alpha} - \Omega'_{\beta\alpha}\Omega'''_{\beta\alpha}} = \Omega_{\alpha\beta}\Omega''_{\alpha\beta} - \Omega'_{\alpha\beta}\Omega'''_{\alpha\beta} \ . \tag{3}$$

Four of these five conditions are independent.

A third time-system π is now introduced. There will be some π-point in π (say p_π) which is occupied by the event-particle (x_a, y_a, z_a, t_a); thus p_π correlates the α-point (x_a, y_a, z_a) with the α-time t_a. Furthermore, it is not difficult to show that in this same way π makes every set of coordinates of a variable α-point a function of t_a, i.e., π correlates a point such as $(x_a + \dot{x}_a dt_a, y_a + \dot{y}_a dt_a, z_a + \dot{z}_a dt_a)$ with the time $t_a + dt_a$. Or, we may say that π correlates an α-point (x_a, y_a, z_a) with the velocity $(\dot{x}_a, \dot{y}_a, \dot{z}_a)$. An analogous conclusion holds for the time-system β. (In intuitive terms, if we think of π as moving through α, the points along the x_a-, y_a-, and z_a-axes appear to be moving with the velocities $dx_a/dt_a, dy_a/dt_a, dz_a/dt_a$, respectively, relative to π.) Also, π correlates the sets of coordinates (x_a, y_a, z_a, t_a) and $(x_\beta, y_\beta, z_\beta, t_\beta)$ with a single event-particle. This permits us to derive from equations (1):

$$\left. \begin{aligned} \dot{x}_\beta &= \frac{\Omega_{\alpha\beta}\dot{x}_a + \Omega'_{\alpha\beta}}{\Omega''_{\alpha\beta} + \Omega'''_{\alpha\beta}\dot{x}_a}, \\[2mm] \dot{y}_\beta &= \frac{\dot{y}_a}{\Omega''_{\alpha\beta} + \Omega'''_{\alpha\beta}\dot{x}_a}, \\[2mm] \dot{z}_\beta &= \frac{\dot{z}_a}{\Omega''_{\alpha\beta} + \Omega'''_{\alpha\beta}\dot{x}_a}. \end{aligned} \right\} \tag{4}$$

Since π is any time-system, it may be identified with α. Then $\dot{x}_a = 0$, $\dot{y}_a = 0$, $\dot{z}_a = 0$; hence $\dot{y}_\beta = \dot{z}_\beta = 0$, and $\dot{x}_\beta$ is the velocity of the time-system α in the space of β. Calling this latter velocity $V_{\beta a}$, we have:

$$V_{\beta a} = \frac{\Omega'_{\alpha\beta}}{\Omega''_{\alpha\beta}}. \tag{5}$$

Identifying π with β, we find

$$V_{\alpha\beta} = -\frac{\Omega'_{\alpha\beta}}{\Omega_{\alpha\beta}}. \tag{6}$$

Assuming that a standard method for choosing the positive directions of the axes OO_{ax} and OO_{at} has been decided upon,[21] the "principle of kinematic sym-

[20] This means that the transformation (1) is the "inverse" or "reciprocal" of the transformation (2), i.e., that the two transformations applied successively to any function of x, y, z, t would leave that function unchanged.

[21] "The positive directions for OO_{at} and $OO_{\beta t}$ are settled by the rule that a positive measure of lapse of time should indicate subsequence in the time-order to the moment O_a. This rule is definite because of the ultimate distinction between antecedence and subsequence in time, which has not otherwise been made use of" (PNK, p. 153).

metry" (which is the seventh axiom for congruence in the present formulation), can be stated as follows:

vii) First, the measures of relative velocities (of β in α and of α in β) are equal and opposite, i.e.,

$$V_{\alpha\beta} + V_{\beta\alpha} = 0;\tag{7}$$

and second, if a velocity of magnitude U in α and normally transverse to the direction of β in α is represented by the velocity $(V_{\beta\alpha}, U')$ in β, where U' is normally transverse to the direction of α in β, then a velocity of magnitude U in β and normally transverse to the direction of α in β is represented by the velocity $(V_{\alpha\beta}, U')$ in α, where U' is normally transverse to the direction of β in α.

The first part of this axiom, according to Whitehead, "may be taken as the definition, or necessary and sufficient test, of such congruence [of the time-units in two time-systems]"; the second part is "the principle of the symmetry of two time-systems in respect to transverse velocities" (PNK, p. 154). The first part of Axiom vii, together with equations (5) and (6), gives:

$$\Omega''_{\alpha\beta} = \Omega_{\alpha\beta}.\tag{8}$$

Now, identifying π with $(\dot{x}_\alpha = 0, \dot{y}_\alpha = U, \dot{z}_\alpha = 0)$, equations (4) and (5) give:

$$\dot{x}_\beta = V_{\beta\alpha}, \qquad \dot{y}_\beta = \frac{U}{\Omega''_{\alpha\beta}}, \qquad \dot{z}_\beta = 0;$$

and identifying π with $(\dot{x}_\beta = 0, \dot{y}_\beta = U, \dot{z}_\beta = 0)$, equations (4) and (6) give:

$$\dot{x}_\alpha = V_{\alpha\beta}, \qquad \dot{y}_\alpha = \frac{U}{\Omega''_{\beta\alpha}}, \qquad \dot{z}_\alpha = 0.$$

Hence, by the second part of Axiom vii,

$$\Omega''_{\alpha\beta} = \Omega''_{\beta\alpha}.\tag{9}$$

Equations (4) can now be written as follows:

$$\left.\begin{aligned} \dot{x}_\beta &= \frac{\Omega_{\alpha\beta}(\dot{x}_\alpha - V_{\alpha\beta})}{\Omega_{\alpha\beta} + \Omega'''_{\alpha\beta}\dot{x}_\alpha}, \\[2mm] \dot{y}_\beta &= \frac{\dot{y}_\alpha}{\Omega_{\alpha\beta} + \Omega'''_{\alpha\beta}\dot{x}_\alpha}, \\[2mm] \dot{z}_\beta &= \frac{\dot{z}_\alpha}{\Omega_{\alpha\beta} + \Omega'''_{\alpha\beta}\dot{x}_\alpha}, \end{aligned}\right\}\tag{10}$$

from equations (4), (6), and (8). Also, from equations (3), (8), and (9):

$$0 = \Omega_{\alpha\beta} - \Omega_{\beta\alpha},$$

$$1 = -\frac{\Omega'_{\alpha\beta}}{\Omega'_{\beta\alpha}} = -\frac{\Omega''_{\alpha\beta}}{\Omega'''_{\beta\alpha}} = \frac{1}{\Omega^2_{\beta\alpha} - \Omega'_{\beta\alpha}\Omega'''_{\beta\alpha}} = \Omega^2_{\alpha\beta} - \Omega'_{\alpha\beta}\Omega'''_{\alpha\beta}.$$

Hence:

$$\Omega'_{\alpha\beta} + \Omega'_{\beta\alpha} = 0, \qquad \Omega'''_{\alpha\beta} + \Omega'''_{\beta\alpha} = 0, \qquad \Omega^2_{\alpha\beta} - \Omega'''_{\alpha\beta} = 1;$$

and by equation (6):

$$\Omega_{\alpha\beta}^2 + V_{\alpha\beta}\Omega_{\alpha\beta}\Omega_{\alpha\beta}''' = 1 \ .$$

Finally, then:

$$\left.\begin{array}{l} V_{\alpha\beta} = - V_{\beta\alpha} \ , \\[1.5ex] \Omega_{\alpha\beta} = \Omega_{\beta\alpha} \ , \\[1.5ex] \Omega_{\alpha\beta}''' = - \Omega_{\beta\alpha}''' \ , \\[1.5ex] \Omega_{\alpha\beta} \left(\Omega_{\alpha\beta} + V_{\alpha\beta}\Omega_{\alpha\beta}'''\right) = 1 \ . \end{array}\right\} \quad (11)$$

Now, $\Omega_{\alpha\beta}$ and $\Omega_{\alpha\beta}'''$ can be expressed in terms of $B_{\alpha\beta}$ and an absolute constant, by assuming the transitivity of congruence with respect to different time-systems:

viii) If the time-unit of α is congruent to the time-unit of β and if the time-unit of β is congruent to the time-unit of γ, then the time-unit of α is congruent to the time-unit of γ.

I shall not reproduce the details of this proof (for which see PNK, pp. 155–57) but simply state the result that if α and β be any pair of time-systems,

$$\frac{V_{\alpha\beta}^2}{1 - \Omega_{\alpha\beta}^{-2}} = k \ , \tag{12}$$

where k is an absolute constant, the same for any such pair α, β.

The possible values of k determine the possible types of kinematics. The value $k = 0$ leads to the result that either $\Omega_{\alpha\beta} = 0$ or $V_{\alpha\beta} = 0$. If $\Omega_{\alpha\beta} = 0$, then, according to equations (1), x_β depends on t_α but not on x_α; while if $V_{\alpha\beta} = 0$, then two time-systems would necessarily be relatively at rest. Both of these consequences are so completely at variance with our experience of motion that the case $k = 0$ need not be considered any further. We turn to the remaining cases with k positive, negative, or infinite, leading respectively to the three types of kinematics which Whitehead calls "hyperbolic," "elliptic," and "parabolic."[22]

Hyperbolic: put c^2 for k. The transformation equations become:

$$\left.\begin{array}{l} x_\beta = \Omega_{\alpha\beta}\left(x_\alpha - V_{\alpha\beta}t_\alpha\right) \ , \\[1.5ex] y_\beta = y_\alpha \ , \\[1.5ex] z_\beta = z_\alpha \ , \\[1.5ex] t_\beta = \Omega_{\beta\alpha}\left(t_\alpha + \dfrac{V_{\beta\alpha}x_\alpha}{c^2}\right), \end{array}\right\} \quad (13)$$

where $\Omega_{\alpha\beta} = (1 - V_{\alpha\beta}^2/c^2)^{-1/2}$.

Also, it follows from equations (13) that

$$x_\alpha^2 + y_\alpha^2 + z_\alpha^2 - c^2 t_\alpha^2 = x_\beta^2 + y_\beta^2 + z_\beta^2 - c^2 t_\beta^2 \ . \tag{14}$$

---

[22] These terms have, of course, nothing to do with hyperbolic, elliptic, and parabolic geometries to which reference was made earlier. In fact, all three types of kinematics presuppose Euclidean (or parabolic) space.

It is easy to show that our previous distinctions among rects, point-tracks, and null-tracks are represented by various values for the invariant (which Whitehead calls *separation*) formulated in equation (14); namely, if we consider the origin O and some other event-particle P, then the separation between O and P will be positive, negative, or zero when the straight route joining O and P is, respectively, a rect, a point-track, or a null-track.

In elliptic and parabolic kinematics, as we shall soon see, null-tracks are non-existent; the special properties of this type of route will therefore be discussed here. Consider two event-particles, P and P', in the coordinate-system a with coordinates, respectively, (x_a, y_a, z_a, t_a) and (x'_a, y'_a, z'_a, t'_a). Suppose now that P and P' are co-momental; then $\Delta t_a = 0$ and the expression $\sqrt{(\Delta x_a^2 + \Delta y_a^2 + \Delta z_a^2 - c^2\Delta t_a^2)}$ represents the "proper" space separation between P and P'. Suppose next that P and P' are sequent; then the expression $\sqrt{(\Delta t_a^2 - [\Delta x_a^2 + \Delta y_a^2 + \Delta z_a^2]/c^2)}$ represents the "proper" time separation between P and P'. Finally, if P and P' are on a null-track, then the expression $\Delta x_a^2 + \Delta y_a^2 + \Delta z_a^2 - c^2\Delta t_a^2$ vanishes and P and P' are termed *co-null*.

Congruence along null-tracks could have been analyzed earlier in connection with the axioms for congruence along rects and point-tracks; it is somewhat simpler, however, to study this special type of congruence in metrical terms, i.e., in terms of the expression for separation characteristic of hyperbolic kinematics. As we have just seen, for any two event-particles on a null-track the separation is zero; when this happens we have $(\Delta x_a^2 + \Delta y_a^2 + \Delta z_a^2)^{1/2} = c\Delta t_a$, and either of these expressions may be taken as the definition of congruence along null-tracks in the single time-system a. On the other hand, neither of the two expressions is invariant with respect to all time-systems—only their difference is invariant and this is always zero for all time-systems. Thus, there is no general definition of congruence along null-tracks which will serve for all time-systems. In the exceptional case of parallel[23] null-tracks, however, a general definition of congruence is possible because stretches along two parallel null-tracks whose respective $c\Delta t$ values are equal in one time-system will have their respective $c\Delta t$ values equal in all time-systems. (Since *only* parallel null-tracks are representable in two-dimensional Minkowski diagrams, such diagrams are seriously misleading in the analysis of congruence along null-tracks.)

Elliptic: put $-h^2$ for k. The transformation equations become:

$$x_\beta = \Omega_{a\beta}\,(x_a - V_{a\beta}t_a)\,,$$

$$y_\beta = y_a\,,$$

$$z_\beta = z_a\,,$$

$$t_\beta = \Omega_{\beta a}\left(-\frac{V_{\beta a}}{h^2}\,x_a + t_a\right),$$

$$\tag{15}$$

[23] "Parallel" here is meant to include oppositely directed null-tracks. Notice that such null-tracks would be represented in a two-dimensional Minkowski diagram (with $c = 1$) by a pair of orthogonal straight lines.

where
$$\Omega_{\alpha\beta} = \left(1 + \frac{V_{\alpha\beta}^2}{h^2}\right)^{-1/2}$$

Also, it follows from equations (15) that

$$x_\alpha^2 + y_\alpha^2 + z_\alpha^2 + h^2 t_\alpha^2 = x_\beta^2 + y_\beta^2 + z_\beta^2 + h^2 t_\beta^2 . \qquad (16)$$

We see from equation (16) that the elliptic formulae have no place for a distinction between different kinds of straight routes between event-particles and that null-tracks in particular are impossible (since the separation given by either side of equation [16] is always positive).

Parabolic: put $k = \infty$. The transformation equations become:

$$\left. \begin{aligned} x_\beta &= x_\alpha - V_{\alpha\beta} t_\alpha , \\[6pt] y_\beta &= y_\alpha , \qquad z_\beta = z_\alpha , \qquad t_\beta = t_\alpha . \end{aligned} \right\} \qquad (17)$$

For the event-particles O and P there are two possible invariant expressions:

$$x_\alpha^2 + y_\alpha^2 + z_\alpha^2 = x_\beta^2 + y_\beta^2 + z_\beta^2 , \qquad \text{for } t_\alpha = t_\beta = 0 ; \qquad (18)$$

and

$$t_\alpha^2 = t_\beta^2 . \qquad (19)$$

Rects here correspond to equation (18) and point-tracks to equation (19); however, in the case of infinite velocities a point-track becomes a rect. Since the invariant expressions in equations (18) and (19) are both always positive (as long as P is distinct from O), we see once more that null-tracks are impossible.

Whitehead now asks which of these three types of kinematics is most closely realized in nature. The parabolic formulae correspond to the Galilean transformation equations of classical mechanics; these formulae are, of course, true to a first approximation (i.e., whenever the relative velocity of two observers, $V_{\alpha\beta}$, is not too large). Both the hyperbolic and the elliptic formulae approximate to the parabolic formulae as the values of h and c approach infinity. However, as we have seen, there is an important difference between the hyperbolic and elliptic formulae: the hyperbolic formulae imply a fundamental distinction between space and time, manifesting itself in the distinction between rects and point-tracks, while no such distinction is implied by the elliptic formulae. But the universal experience of mankind testifies to a sharp distinction between space and time, so that the elliptic formulae must be rejected.

Thus we are left with the hyperbolic formulae; these are identified with "the formulae of the Larmor-Lorentz-Einstein theory of electromagnetic relativity, namely, the theory by which with a certain amount of interpretation the electromagnetic equations are invariant for these transformations" (PNK, p. 159). Also, Whitehead says, "experiment has, so far, pronounced in favour of the hyperbolic type [of formulae]" (PNK, p. 164). However, Whitehead emphasizes the fact that the assumption of the invariance of the electromagnetic equations (or, what amounts to the same thing, the invariance of the speed of light) has played no role in his deduction of hyperbolic kinematics. Thus, Whitehead's interpretation of the significance of the universal constant c which occurs in the formulae of hyperbolic kinematics is that it represents a velocity whose magnitude is the same in all time-systems and which consequently represents the upper bound to the possible velocities of time-systems relative to one another. That c is closely approximated by, if not identical with, the velocity of electromagnetic waves *in vacuo* is for Whitehead a contingent characteristic of our physical universe (perhaps only during the present cosmic epoch?). The identification of c with the velocity of these waves enables us to give a definite physical interpretation to the null-tracks of our earlier discussions (namely, null-tracks represent possible paths of light-rays *in vacuo*) and also enables us to use hyperbolic kinematics to explain all of the results usually included in special relativity kinematics. Furthermore, the geometry of the space-time of hyperbolic kinematics is identical with Minkowskian pseudo-Euclidean geometry; this is evident from equation (14), which expresses the invariance of interval. Whitehead, however, does not introduce the concept of imaginary time; this rather minor mathematical detail stems, I believe, from Whitehead's intention of making as plain as possible the place of *perceived* space and time in physical theory.[24]

Null-tracks, as we have already noted, play a central role in the space-time manifold of hyperbolic kinematics, namely, the two three-dimensional "conical" surfaces of null-tracks which pass through any given event-particle E serve to subdivide the set of all other event-particles into three mutually exclusive sets: (1) past with respect to E, (2) future with respect to E, and (3) co-present with respect to E. Co-presence, as already noted, is a symmetrical but a non-transitive (i.e., neither transitive nor intransitive) relation. Whitehead comments:

> The properties of co-present event-particles are undeniably paradoxical. We have, however, to remember that these paradoxes occur in connexion with the ultimate baffling mystery of nature—its advance from the past to the future through the medium of the present [ET, p. 246].[25]

Although Whitehead admits that velocities greater than c must have a physical significance "entirely different from" velocities less than c (PNK, p. 160), he never asserts (as most orthodox relativists do) that velocities greater than c are entirely devoid of physical significance. As a matter of fact, far from accepting

[24] Cf. PNK, p. vi: "The whole investigation is based on the principle that the scientific concepts of space and time are the first outcome of the simplest generalisations from experience, and that they are not to be looked for at the tail end of a welter of differential equations."

[25] In later works, Whitehead uses the term "contemporaneous" for the relation here designated as "co-present." Whitehead's final discussion of the ontological and ethical implications of this relation occurs in AI, chap. xii. For a brief account of the ontological implications see below, p. 144. The ethical implications center about the fact that "The causal independence of contemporary occasions is the ground for the freedom within the Universe" (AI, p. 255).

the limiting character of the velocity of light as an ultimate fact of nature, Whitehead is willing to entertain seriously an objection to the hyperbolic formulae based on the mere existence, implied by these formulae, of an absolute velocity c. Whitehead says:

In the hyperbolic kinematics there is an absolute velocity c with special properties in nature. The difficulty which is thus occasioned is rather an offence to philosophic instincts than a logical puzzle. But certainly our familiar experience, in some way which it is difficult to formulate in words, leads us to shun the introduction of such absolute physical quantities [PNK, p. 164].

The difficulty is mitigated by the realization that the existence of the absolute velocity c really means that space and time are comparable manifolds with respect to congruence; or, in other words, that there is a "natural relation" between space-units and time-units which can be expressed by making c unity (PNK, p. 164). Since this comparability of space-units and time-units is only a contingent feature of the physical world, there is no necessity for expecting its manifestation in ordinary sense-experience. In fact, it requires extremely careful and precise measurements to detect any phenomenon associated with the presence of c in hyperbolic kinematics, and this serves to account for the seeming paradox of spatio-temporal congruence.

Some of the more important differences in Einstein's and Whitehead's respective derivations of the Lorentz equations may now be discussed. The objectives of Whitehead's philosophy of nature determine his approach, in particular his refusal to appeal to such special physical data as the existence of a unique physical process (electromagnetic radiation) with a velocity whose magnitude is a universal constant. In Einstein's derivation of the Lorentz equations, it will be recalled, the definition of simultaneity, hence ultimately of time-congruence, depends upon a postulate as to the constant velocity of light; for Whitehead, on the other hand, the behavior of light is a matter of theoretical prediction from the laws of the electromagnetic field rather than a matter of ultimate brute fact to be postulated without explanation (see below, pp. 205–6). Whitehead's eschewal of such postulates forces him to define time-congruence by an alternative procedure involving only assumptions about the *general* perceived character of motion (the principle of kinematic symmetry and the transitivity of time-congruence). Thus one symptomatic difference in Einstein's and Whitehead's derivations of the Lorentz equations occurs in the contrasting logical roles of the principle of the equality of relative velocity for two observers in uniform relative motion: for Whitehead this principle is an axiom exemplifying a fundamental directly observable symmetry-property of motion; for Einstein this principle is a theorem—an immediate consequence of the Lorentz equations themselves.[26]

The second of Einstein's two basic postulates—the principle of (special)

[26] Other writers have derived the Lorentz equations by taking the equality of relative velocity as an axiom; see, e.g., E. T. Whittaker, *From Euclid to Eddington*, p. 61: ". . . each system regards the other as moving with relative velocity w."

relativity—is also repugnant to Whitehead's general philosophy of nature. From Whitehead's point of view Einstein's relativity principle is illegitimately a priori, since "It is not at all obvious that invariance of form in respect to all time-systems is a requisite in the complete expression of such laws; namely, the demand for relativistic equations is only of limited applicability"[27] (PNK, p. 161).

12. Congruence and Measurement

To attain full clarity about the relation of congruence as it functions in the natural sciences, one must, according to Whitehead, view it as a subspecies of the more general relation of *equality*.[28] Here equality does not refer necessarily to quantitative equality, and of course equality must not be confused with identity. Equality, in its important applications, always implies a diversity of things related and an identity of character (e.g., two physical objects which "match" in color). The following notation serves to keep these fundamental distinctions before us.

Let γ, the *qualifying class*, represent a class of characters, which may be finite, denumerably infinite, or non-denumerably infinite. We assume that there is a (finite or infinite) class of things A, B, C, . . . , each of which possesses exactly one character of the class γ; this class of things is called the *qualified class for* γ. Then,

$$A = B \rightarrow \gamma$$

means that the same member of γ qualifies both A and B (or, A and B are equal with respect to the class γ); and

$$A \neq B \rightarrow \gamma$$

means that one member of γ qualifies A and another member of γ qualifies B (or, A and B are unequal with respect to the class γ). It follows that

$$A = B \rightarrow \gamma \quad \text{and } B = C \rightarrow \gamma \quad \text{imply } A = C \rightarrow \gamma ,$$

[27] Cf. CN, pp. 196–97: "It has been laid down that these laws [of nature] are to be expressed in differential equations which, as expressed in any general system of measurement, should bear no reference to any other particular measure-system. This requirement is purely arbitrary. For a measure-system measures something inherent in nature; otherwise it has no connexion with nature at all. And that something which is measured by a particular measure-system may have a special relation to the phenomenon whose law is being formulated. For example the gravitational field due to a material object at rest in a certain time-system may be expected to exhibit in its formulation particular reference to spatial and temporal quantities of that time-system. The field can of course be expressed in any measure-systems, but the particular reference will remain as the simple physical explanation."

I shall return to this point in connection with Whitehead's theory of relativity (see below, pp. 208, 212).

[28] The following discussion is based mainly on R, chap. iii, Whitehead's most detailed treatment of the subject.

which, as Whitehead says, is a general rendering of Euclid's first axiom ("Things equal to the same thing are equal to each other"). It should be noted that $A = B \rightarrow \gamma$ and $A \neq B \rightarrow \gamma$ are contraries, not contradictories, so that both may be false (e.g., when A or B or both possess no character or more than one character from γ).

By laying down an appropriate set of axioms for the equality relation, the more special relation of congruence may be specified. (The axioms which Whitehead introduces at this point hold for both space and time; accordingly I shall speak of the congruence relation for a "manifold," using this latter term instead of Whitehead's term "space.") We consider the simplest possible case, that of a one-dimensional manifold whose points are arranged by some definite relation in serial order with Dedekindian continuity, infinite in both directions.[29] Then, the qualified class for γ is taken to consist of all the finite stretches A, B, C, . . . , in the manifold; while, in order for $A = B \rightarrow \gamma$ and $A \neq B \rightarrow \gamma$ to specify a congruence relation, the qualifying class γ must be a class of qualities (usually called "magnitudes"), exactly one of these qualities characterizing each finite stretch and such that all the members of γ can be described in terms of one of them (the "unit") by means of real numbers. Whitehead lists eight axioms which suffice to make γ satisfy the required conditions for a "congruence class." Each of these axioms expresses a familiar condition so that it will suffice here to express the essential point of each axiom by a rough verbal formulation (see R, pp. 46–48).

i) From a given point in the manifold, stretches of an assigned magnitude can be measured in either direction.

ii) A stretch is unequal to any of its (proper) parts.

iii) Equal stretches added to equal stretches give equal stretches.

iv) Equal stretches subtracted from equal stretches give equal stretches.

v) It is never the case that both A equals a part of B and B equals a part of A.

vi) If A be any stretch and n any integer, then a sequence of n equal stretches which add up to A can always be found.

vii) If A and B be two stretches with a common end-point and A be part of B, then an integer n can always be found such that there exists a sequence of n equal stretches of which A is the first, whose sum is greater than B. (This, the so-called "Archimedian axiom," guarantees that any given point in the manifold can be reached by successive addition of a finite number of equal stretches.)

viii) If A is any stretch and n any integer, then A is a member in any assigned ordinal position (less than or equal to n) of two sequences of n equal stretches each, running in opposite directions.

Whitehead stresses the fact that

. . . a quality which belongs to the set γ is in itself in no way otherwise distinguished from any other quality of things. Quantity arises from a distribution of qualities which

---

[29] The concept of a continuous, n-dimensional manifold of points is undefined here; in other contexts it may be desirable to define the concept.

in a certain definite way has regard to the peculiar fact that in certain cases two extended spatio-temporal elements together form a third such element. In fact the 'qualifying' qualities are distributed among extended things with a certain regard to their property of extension [R, p. 49].

From this it follows that stretches equal for one qualifying class γ may be unequal for another qualifying class γ'. This conclusion will be illustrated in the discussion below.

Suppose that some method has been devised for specifying the congruence of all stretches in a given one-dimensional continuum (which may conveniently be called a "line"). This means that a real number has been correlated with each stretch on the line in accordance with the above axioms. There will certainly be an infinite number of ways of effecting this correlation, but all of them are essentially equivalent to one another (i.e., in more precise terms, all topologically isomorphic lines are also metrically isomorphic). Another way of putting this is to say that any line (ignoring topological properties such as whether the line is open or closed) can be bent without stretching into coincidence with any other line so that all congruence relationships are preserved. In the two-dimensional case this is obviously no longer true; thus a portion of a sphere's surface (unlike, in this respect, a portion of a cylinder's or a cone's surface) cannot be bent without distortion, and consequent loss of congruence relationships, into coincidence with a Euclidean plane. The sphere and the plane possess an *intrinsic* curvature for a given metrization; the circle and the straight line do not.[30] Consequently, since different portions of a given surface may possess (in an undefined, intuitive sense) different "degrees of curvature," it will not generally be possible to test figures in the surface for congruence by attempting to slide them *in the surface* into superposition. Intuitively, then, it would seem that if congruence is to be defined so as to apply to all finite figures in a given surface, then that surface must be of *constant curvature* (e.g., a spherical surface). This result (generalized to three dimensions) constitutes the solution to the "Riemann-Helmholtz problem" of determining all those spaces in which "free mobility" is possible; it was first rigorously proven by Sophus Lie, whose method was to give a precise characterization of the possible motions of a rigid body and then to deduce the meanings of congruence and hence the types of metrical

---

[30] However, a meaning can be given to one-dimensional Euclidean or non-Euclidean geometries. We begin with a straight line in the Euclidean plane, and then define non-Euclidean distances along this line by the areas of sectors (in the Euclidean plane) included between the line and an ellipse or hyperbola tangent to the line. It turns out that if the Euclidean metric on the line is denoted by dx, then the elliptic metric can be written as: $dx_e = dx/(1 + x^2)$; and the hyperbolic metric as: $dx_h = dx/(1 - x^2)$. These two expressions are special cases of the general Riemannian expression for metric in a continuous manifold of constant curvature (see below, p. 130); for further details see H. S. M. Coxeter, *Non-Euclidean Geometry*, pp. 253–54, 265. That a line possesses no *intrinsic* curvature is here indicated by the fact that one must begin by simply *assuming* that a certain line (namely, a straight line in a Euclidean plane) is "Euclidean." The significance of this result for the one-dimensional time continuum is discussed below, p. 92.

geometry compatible with such motions. The metrical geometries to which Lie was led are the ordinary Euclidean type (also called parabolic) and the two non-Euclidean types: Bolyai-Lobatchewskian (or hyperbolic) and Riemannian (or elliptic).[31]

One further aspect of Lie's analysis of congruence should be noted: corresponding to each of the three metrical geometries there is not one but an infinite number of different meanings of congruence. In order to illustrate this point, let us assume that the metrization of an ordinary physically "flat" two-dimensional surface S is effected by introducing a rectangular Cartesian coordinate-system x-y on S, so that the distance between any two points on S is given by the expression:

$$d s^2 = d x^2 + d y^2 ,$$

i.e., the Pythagorean formula for two-dimensional Euclidean geometry. Suppose, however, that we substitute for ds^2 the expression:

$$d s'^2 = a^2 d x^2 + d y^2 \qquad (a = \text{constant}) , \quad (20)$$

and consider the relative lengths of a pair of segments parallel respectively to the two coordinate axes. If these two segments are congruent according to the ds^2 definition, they will *not* be congruent according to the ds'^2 definition (the segment parallel to the x-axis will equal a times the other segment). But it can be shown that the ds'^2 definition of distance also leads to Euclidean geometry on S.[32] What the ds'^2 definition in effect does is to make the length of a standard measuring rod depend on its orientation in S.

We are now in a position to understand the philosophical significance which Whitehead attributes to Lie's results in the following passage:

If we apply . . . [the above account of congruence] to the classical theory of space and time, we find, following Sophus Lie's analysis, that there are an indefinite number of qualifying classes $\gamma, \gamma', \gamma''$, etc., which for the case of three-dimensional space generate relations of congruence among spatial elements, and that each such set of congruence relations is inconsistent with any other such set.

For the case of time the opposite trouble arises. Time in itself, according to the classical theory, presents us with no qualifying class at all on which a theory of congruence can be founded.

This breakdown of the uniqueness of congruence for space and of its very existence for time is to be contrasted with the fact that mankind does in truth agree on a congruence system for space and on a congruence system for time which are founded on the direct evidence of its senses. We ask, why this pathetic trust in the yard-measure and the clock? The truth is that we have observed something which the classical theory does not explain [R, p. 49].

---

[31] Two-dimensional elliptic geometry is illustrated by the surface of a sphere; there is no familiar surface in Euclidean space which illustrates two-dimensional hyperbolic geometry. However, a hyperbolic surface without constant curvature is illustrated by the surface of a saddle.

[32] In fact, any expression for ds'^2 with constant coefficients for dx^2 and dy^2 leads to Euclidean geometry on the surface (see below, pp. 129–30).

The "classical theory" of space and time refers here, one must assume, to Newton's absolute space and absolute time, each of which was conceived as a perfectly definite continuous manifold, existing quite independently of its "contents," and with perfectly definite metrical characteristics. In the light of subsequent geometrical developments, however, the classical theory became untenable. For, on the one hand, time, being a one-dimensional continuum, does not possess any intrinsic metrical characteristic. (This does not imply, of course, that a metric cannot be imposed, so to speak, "from without" on the temporal continuum. One such procedure for metrizing time would be to assume the validity of Newton's laws of motion in their usual mathematical formulation, thereby specifying a definite meaning for temporal congruence, i.e., those time-lapses are congruent which, when they are assigned equal numerical values, are consistent with numerical values for acceleration satisfying the formula $F = ma$. Other metrizations for time incompatible with the "Newtonian" time just characterized can be obtained by taking non-linear functions of the "Newtonian" time. That this procedure for metrizing time is consistent with Newton's concept of absolute time I seriously doubt, but this is of no great importance from the Whiteheadian point of view since, as we shall see, Whitehead finds the very procedure itself paradoxical in the extreme.) Space, on the other hand, being a three-dimensional continuum, is in itself open to intrinsically different types of metrical determination (each of the three main types corresponding to an infinity of alternative definitions of congruence). But, with respect to absolute space, it is difficult to see what criterion could be used to single out the "true" metrical geometry and the "true" definition of spatial congruence.

What now of modern theories of space and time; do they fare any better than the classical theory in their accounts of temporal and spatial congruence? According to Whitehead, those modern theories which share with the classical theory the root conception of nature at an instant as an independent fact must also fail to explain congruence adequately.[33] That such modern theories often tend to be of a "conventionalist" cast is from the Whiteheadian point of view presumably no mere accident: by retaining certain central features of the materialist philosophy of nature, these theories are (like the classical theory) compelled to ignore the genuine factors in nature which underlie spatial and temporal congruence; and, since absolute space and time are no longer acceptable, the one remaining alternative seems to be conventionalism.

Now, Whitehead is perfectly willing to accept one part of the conventionalist thesis with respect to geometry, namely, that if any one of the three geometries with uniform curvature is exemplified in a given region of the physical world, then each of the other two geometries is also exemplified in that region. This may be illustrated as follows. Suppose we begin, as above (p. 91), with the

---

[33] E.g., the rival interpretations of the epistemological status of geometry espoused by Poincaré and Bertrand Russell (see CN, pp. 122–23).

"flat" physical surface S metrized by the Pythagorean formula for distance. By introducing as a new definition of distance the expression

$$d s^2 = \frac{d x^2 + d y^2}{K y^2} \qquad (K = \text{a constant} > 0) , \quad (21)$$

we obtain a representation (known as the "Poincaré half-plane") on S—or at least on that part of it for which $y > 0$—of the hyperbolic plane with constant curvature $-K$.[34] The preceding result, however, Whitehead considers of purely mathematical interest:

For in the three metrical geometries as applying to the same subject-matter, the definitions of distance are different. . . . There are three diverse systems of relationship within the subject-matter, so related that if one be present then the others are present. Also, of course, the description of one set of relationships can be achieved, though very clumsily, in terms of any one of the other two sets [e.g., the spatial relations on the surface of a sphere may be neatly characterized by spherical geometry, only clumsily by Euclidean geometry]. There is nothing 'conventional' in this, except the obvious fact that we can direct attention to any selected group of facts [AI, pp. 174–75].

Likewise of purely mathematical interest is the previously discussed multiplicity of possible definitions of distance (and hence of congruence) corresponding to a given type of metrical geometry. According to Whitehead, everyone in fact does agree, "apart from minor inaccuracies of perception," upon a single meaning for distance. Thus, "It is a fact of nature that a distance of thirty miles is a long walk for any one. There is no convention about that" (AI, p. 175). But precisely what is it in nature that grounds this agreement on the meaning of distance and congruence? Whitehead's answer, as has been suggested above, is an appeal to that definite factor in sense-awareness which consists in our ability to recognize persistence of quality throughout the time-span of a single specious present:

Congruence is a particular example of the fundamental fact of recognition. In perception we recognise. This recognition does not merely concern the comparison of a factor of nature posited by memory with a factor posited by immediate sense-awareness. Recognition takes place within the present without any intervention of pure memory. For the present fact is a duration with its antecedent and consequent durations which are parts of itself [CN, p. 124].

Thus, the theory of congruence is one branch of *the theory of recognitions* (another branch is the theory of objects, to be discussed in chap. vii).[35]

[34] On K, the so-called "measure of curvature," see below, p. 129. On the Poincaré half-plane, see H. Poincaré, *Science and Hypothesis*, pp. 59–60.

[35] In the most general sense of the term "object"—namely, anything recognizable—the relation of extension (and hence presumably all extensive concepts such as congruence) is itself an object. However, Whitehead prefers to "restrict the term ["object"] to those objects which can in some sense or other be said to have a situation in an event . . ." (CN, p. 189), thereby excluding extension and congruence.

What is recognized in any perception of congruence is always the constancy of some objects as they appear (or "ingress," to use Whitehead's technical term) in several simultaneous or successive events. The objects whose constancy is recognized in the perception of congruence will usually include material objects (such as a measuring rod or a pendulum) as well as certain sets of physical conditions (such as the "uniformity of conditions for the uniform transmission of light" which, Whitehead holds, is presupposed by Einstein's definition of simultaneity—cf. PNK, p. 55). The general geometrical principle exemplified in all perceptions of congruence can be stated as follows: "Two segments are congruent when there is a certain analogy between their functions in a systematic pattern of straight lines, which includes both of them" (PR, p. 505).

Judgments of constancy based on direct recognition within sense-awareness are neither infallible nor incorrigible; nevertheless, "though isolated judgments may be rejected, it is essential that a rational consideration of nature should assume the truth of the greater part of such judgments and should issue in theories which embody them" (PNK, p. 56). Once some immediate judgments of congruence are available, inferences as to the existence of other not directly observable congruence-relations become possible; in fact, it is precisely for the sake of facilitating such inferences that the various procedures of measurement have been devised. But the very measuring procedures themselves always presuppose immediate judgments of congruence: "all experimental measurement involves ultimate intuitions of congruence between earlier and later states of the instruments employed" (PR, p. 508). The most obvious example is that of a measuring rod made of some "rigid" material: we distinguish such a rod from one made of an "elastic" material (and therefore unsuitable for measurement of length) by direct judgments of the self-congruence of the rigid rod as we lay it off along the stretch being measured. It is true that in practice we may attempt to correct the readings of the measuring rod for the effects of temperature, gravity, etc., rather than relying on direct perceptions of self-congruence from one instance of the use of the measuring rod to another; on the other hand, even such corrections depend in the last analysis on direct judgments of constancy. As Whitehead says:

> However far the testing of instruments and the corrections for changes of physical factors, such as temperature, are carried, there is always a final dependence upon direct intuitions that relevant circumstances are unchanged. Instruments are used from minute to minute, from hour to hour, and from day to day, with the sole guarantee of antecedent tests and of the *appearance* of invariability of relevant circumstances [PR, p. 502].

The meaning of temporal congruence poses a special problem for a materialistic philosophy of nature, since, according to its tenets, no direct inspection of the "coincidence" of time-lapses is ever possible. The usual way out of this difficulty is to introduce a further dose of that conventionalism which, it is alleged, is already required for the specification of spatial congruence: in analogy to the *convention* that one's standard measuring rods are permanently self-congruent there is the *convention* that one's standard clocks always run uniformly. Furthermore, we are told, the "appropriateness" or "convenience" (not "truth"!) of the assumption that one's clocks run uniformly is based on the fact that this assumption is consistent with the usual (mathematically simple) formulation of Newton's laws of motion. Whitehead's criticism of the conventionalist account of temporal congruence is simply that it is unbelievable; mankind did not have to wait for the discovery of the laws of motion before it could recognize equal lapses of time.[36] The conventionalist appeal to "molecular clocks" is no more convincing:

I expect to be told by some that the comparison of equal times is a convention, founded upon an arbitrary selection of a certain type of recurring phenomena as periodic in equal lapses of time. For example, molecules will be brought into evidence, and we shall then found ourselves on Einstein's dictum that a molecule is a natural clock. Let it be noted that on this theory the assumption is purely arbitrary: there can be no sense in saying that it is nearly true and approximately verified, for the very *meaning* of equality of time-lapses is involved. But if this be the case, the explanation of the identity of colours of light emitted by molecules of the same type, as being due to vibrations in equal periods, becomes mere nonsense. For it cannot be due to the fact that we *call* the times equal [PS, p. 39].

I shall now summarize Whitehead's theory of congruence, attempting— what Whitehead himself never quite does—to relate the general doctrines of his natural philosophy to his formal axioms for congruence. First of all, sense-awareness by relatedness indicates the uniformity of spatio-temporal relations. This implies that the three-dimensional geometry of space must be parabolic (Euclidean), elliptic, or hyperbolic. No similar result follows for time, since there is nothing corresponding to Euclidean or non-Euclidean character in a one-dimensional continuum. Sense-awareness by adjective now serves to single out the particular meanings for spatial and temporal congruence appropriate to a given time-system, i.e., to a given meaning of rest and motion within nature —which meaning, it will be recalled, is specified by the relation of cogredience between a percipient event and its entire simultaneous environment. Here is where the axioms for congruence come in: they apply primarily to the adjectives (or objects) recognized in sense-awareness. The axioms are designed to extend the meaning of congruence beyond those instances of congruence immediately recognizable as such in sense-awareness. Thus, in the case of time we are, for example, directly aware within a single specious present of the congruence of the two successive time-intervals between three consecutive $\frac{1}{5}$-second jumps

[36] Cf. CN, p. 137: "King Alfred the Great was ignorant of the laws of motion, but knew very well what he meant by the measurement of time, and achieved his purpose by means of burning candles. Also no one in past ages justified the use of sand in hour-glasses by saying that some centuries later interesting laws of motion would be discovered which would give a meaning to the statement that the sand was emptied from the bulbs in equal times. Uniformity in change is directly perceived, and it follows that mankind perceives in nature factors from which a theory of temporal congruence can be formed. The prevalent theory entirely fails to produce such factors."

of the second-hand on a watch (the duration of the specious present would then be about $\frac{2}{5}$ second, not an unreasonable value). The axioms for temporal congruence then enable us to extend this meaning of temporal congruence at a certain point within the space of a given time-system to any other point within that space and eventually to the spaces of all other time-systems. Similarly, in the case of space, we are directly aware within a single specious present of the continued self-congruence of a rigid rod. The axioms for spatial congruence then enable us to extend this meaning of spatial congruence at a certain point within the space of a given time-system to any other point within that space and eventually to the spaces of all other time-systems.

Let us now examine a few possible applications of the axioms for congruence. Axiom i says that the opposite sides of a parallelogram are congruent. These sides, we recall, must be either both rects (spatial distances) or both point-tracks (temporal lapses). But all point-tracks associated with a given time-system are parallel, which simply means that the time-lapse between events at a certain fixed point in the space of a given time-system and the time-lapse between events at some other fixed point in the same space will always be represented by parallel lines. Under what circumstances, we want to know, will routes along these two parallel point-tracks be equal and hence the time-lapses congruent? The answer is, according to Axiom i, when the rect joining the initial events and the rect joining the final events of the time-lapses are also parallel. One circumstance in which this latter condition is satisfied is when the two initial events occur (ideally) at the same moment and likewise for the two final events. Suppose, for example, that I am simultaneously looking at the second-hand of my watch and listening to the chimes in a distant steeple. I perceive directly that one particular sound of the chimes (which, according to Whitehead, I hear not just "in my head" but also and literally *in the steeple*) is simultaneous with one particular position of the second-hand on the dial; and I make the same observation for another sound of the chimes and another position of the second-hand. Now, a naïve (not necessarily even conscious) application of Axiom i would lead me to infer that the time-lapse registered on my watch is congruent with the time-lapse between the two sounds which I hear in the steeple. But since the sounds seem to be caused by two physical chimings also in the steeple, I naturally infer that the time-lapse registered on my watch is also congruent with the time-lapse between these two physical chimings. (It is assumed that I observe the stationariness, within the space in which I am at rest, of the steeple in which I hear the chimes sounding.) Such naïve experiences, according to Whitehead, form the very foundation of all knowledge of the natural world, including scientific knowledge. But, of course, I may decide that my inference is erroneous (if, for example, it contradicts the observations of someone stationed in the steeple). I may learn, in other words, that there are relevant conditions which must be satisfied in order for Axiom i to be validly applied (e.g., any wind between me and the steeple must be of constant velocity during the course of my observations). My belief that the relevant conditions are, or are not, satisfied is to a first approximation also based on direct sense-perception.

It must not be thought that the application of Axiom i is confined to cases in which part of what one observes is something in one's immediate vicinity (such as my watch). When an astronomer looks at the night sky he sees a set of simultaneous star-images; when he looks again a few minutes later he sees another set of simultaneous star-images. For the purpose of interpreting what he sees (which might just as well be recorded on two different photographic plates) he necessarily assumes that the time-lapse between earlier and later images of the same star is equal for all stars. Without this assumption it would make no sense to trace the motion of some object such as a planet against the background of "fixed stars." Nothing, to be sure, directly follows from this assumption as to the congruence of time-lapses between star-images concerning the possible temporal and causal relations of the stars themselves. On the other hand, such relations among stars—which it is the goal of astronomy to discover—can only be studied by means of complex inferences based ultimately on patterns of observed star-images. The fundamental fact presupposed in such inferences is that the conditions affecting the transmission of light from the stars to the terrestrial observer are essentially the same on the two occasions when the sky is observed. The fact of this constancy of relevant conditions in the space between the stars and the observer is attested, once again, in the first instance by direct sense-perception.

Temporal congruence of time-lapses in different time-systems is defined in terms of the equality of relative velocity for the two time-systems (Axiom vii). This is straightforward enough (remembering, of course, that velocity presupposes spatial as well as temporal congruence within a given time-system). Before turning to spatial congruence, however, we must clarify one further aspect of temporal congruence. Reference has been made repeatedly to the "space" of a given time-system; but which space is this—instantaneous or time-less? Obviously, the latter is required for practical and scientific purposes, and just as obviously, restriction of directly observed temporal congruence to the confines of a specious present inevitably restricts the meaning of temporal congruence to (more or less) a single instantaneous space. How can the transition from instantaneous spaces to time-less space be made? The answer is, by directly observing that the relevant characteristics of one's "clock" as well as the relevant environmental conditions remain sufficiently constant to insure the clock's uniformity—one more crucial instance of the recognition of permanence in sense-awareness by adjective.

And a similar recognition of permanence is required in the transition from spatial congruence in instantaneous spaces to spatial congruence in time-less space: here one must directly observe the presence of the relevant conditions required for one's measuring rods to remain self-congruent. The application of the axioms for spatial congruence involves no novel principles—simply the metrical theorems of Euclidean geometry—but it is presupposed that we can both directly perceive and also construct parallel and perpendicular spatial stretches. For example, we may recognize at a glance that one of a certain pair of rects is a perpendicular bisector of the other, and hence (applying Axiom iv) that the two rects constitute respectively the altitude and the base of an iso-sceles triangle. Whitehead is, of course, assuming that the repeated use of these Euclidean axioms for indirectly determining spatial congruence has never led to inconsistencies with our direct judgments of spatial congruence. As science develops, hitherto inaccessible types of events become susceptible of such in-direct determinations of congruence (e.g., events in atoms and events highly remote in space and time), but always the direct judgments of congruence in sense-perception remain fundamental. It is these latter judgments of congruence that Whitehead has in mind when he criticizes the conventionalist view of con-gruence in such passages as the following:

> The measurement of time was known to all civilised nations long before the laws [of motion] were thought of. It is this time as thus measured that the laws are con-cerned with. Also they deal with the space of our daily life. When we approach to an accuracy of measurement beyond that of observation, adjustment is allowable. But within the limits of observation we know what we mean when we speak of measure-ments of space and measurements of time and uniformity of change. It is for science to give an intellectual account of what is so evident in sense-awareness. It is to me thoroughly incredible that the ultimate fact beyond which there is no deeper explana-tion is that mankind has really been swayed by an unconscious desire to satisfy the mathematical formulae which we call the Laws of Motion, formulae completely unknown till the seventeenth century of our epoch [CN, p. 140].

In other words, Whitehead is arguing that the near-universal acceptance of certain standards of length (e.g., rods made of special kinds of material) and of time (e.g., celestial motions) is compelling evidence that some perceptions of spatial congruence and of temporal congruence are identified as providing the true basis for all space and time measurements. Not all such perceptions are valid, since various physiological and psychological factors may interfere, but *some* ultimate basis in *some* set of sense-perceptions is necessary.

The time and space "of our daily life" in the above quotation can hardly refer to the Minkowski space-time of physics, since the defining characteristic of this latter manifold (namely, the expression for pseudo-Euclidean metric) is only pertinent to the interpretation of some rather special and delicate experiments. On the other hand, there is no real contradiction between commonsense space and time and the space-time of theoretical physics. And just as Whitehead adopts Minkowski space-time for the formulation of his theory of relativity, so he is ready to substitute hyperbolic or elliptic for Euclidean space in that formu-lation "if any observations are more simply explained by such a hypothesis" (R, p. v). Since the difference between Euclidean space and hyperbolic or elliptic space is a matter of degree, one might consistently hold that within the immedi-ate region of directly observable spatial relations space is practically Euclidean, whereas in more extended regions space is significantly "curved."

It is worth remarking that Whitehead's interpretation of the epistemological status of geometry differs in just one respect—but a crucial one—from such recent variants of the conventionalist position as those set forth by Reichenbach and Grünbaum.[37] Whitehead agrees with these philosophers that the question of the nature of physical space is open to experimental test once a definite meaning for congruence has been adopted. However, Whitehead insists that the meaning for congruence which we adopt is not a matter of free choice but rather that a particular meaning is forced on us by the deliverances of sense-awareness, and that this meaning of congruence is necessarily such as to lead to a uniform structure for space. This doctrine concerning the nature of geometry Whitehead had defended at least as early as 1910, the date of the following passage, which was written as a direct reply to Poincaré's conventionalism.

This point of view [i.e., Poincaré's] seems to neglect the consideration that science is to be relevant to the definite perceiving minds of men; and that (neglecting the ambiguity introduced by the invariable slight inexactness of observation which is not relevant to this special doctrine) we have, in fact, presented to our senses a definite set of transformations forming a congruence-group, resulting in a set of measure relations which are in no respect arbitrary. Accordingly our scientific laws are to be stated relevantly to that particular congruence-group. Thus the investigation of the type (elliptic, hyperbolic or parabolic) of this special congruence-group is a perfectly definite problem, to be decided by experiment [AG, p. 192].

13. The Philosophical Import of the Method of Extensive Abstraction

At the beginning of my analysis of Whitehead's philosophy of nature I formulated two leading questions (p. 22 above): (1) What things are observed? (2) What is an observed character? The answers to these questions involve the philosophical import of the constants of externality (already discussed in chapter iv), of objects (to be discussed in chapter vii), and of the method of extensive abstraction.

In the course of his development of the method of extensive abstraction (culminating in the derivation of the Lorentz equations) Whitehead assumes, explicitly or implicitly, several sets of axioms: axioms for extension, axioms on the continuity of events, axioms of serial order for the moments of any time-system and for the event-particles on a route, axioms for cogredience, and axioms of congruence for space and time. What is the justification of, or evidence for, these axioms? In answering this question we may take as our point of departure the following statement by Whitehead:

The importance of this procedure [the method of extensive abstraction] depends on certain properties of extension which are laws of nature depending on empirical verification. There is, so far as I know, no reason why they should be so, except that they are [PNK, p. 75].

[37] See, for example, H. Reichenbach, *The Philosophy of Space and Time*, and A. Grünbaum, "Conventionalism in Geometry."

Now, all of Whitehead's axioms are stated in terms of the constants of externality and concepts defined by means of these constants. Hence, these axioms should be, in principle, as confirmable by direct perception as the constants of externality are discoverable by direct perception. In fact, from what Whitehead says of these constants, it would seem that they are known only insofar as their mutual interrelations (and therefore the axioms) are known. Also, it is clear that the axioms cannot be directly confirmed by physical measurement, since the axioms are in all respects prior to actual measuring procedures. The proper interpretation of Whitehead's assertion that his axioms are "laws of nature depending on empirical verification" is that there are certain elements together with their interrelations given in perception, of which these axioms are the idealized and generalized counterparts. Each of the axioms is *sometimes* clearly exemplified in our experience, and its mode of exemplification leads us to postulate its validity throughout all possible experience. So much for the axioms of the method of extensive abstraction; now let us examine the method itself.

I begin with a passage in which Whitehead analogizes the method of extensive abstraction to the theory of limits in the differential calculus.

It [the method of extensive abstraction] is a method which in its sphere achieves the same object as does the differential calculus in the region of numerical calculation, namely it converts a process of approximation into an instrument of exact thought. The method is merely the systematisation of the instinctive procedure of habitual experience. The approximate procedure of ordinary life is to seek simplicity of relations among events by the consideration of events sufficiently restricted in extension both as to space and as to time; the events are then 'small enough.' The procedure of the method of extensive abstraction is to formulate the law by which the approximation is achieved and can be indefinitely continued. The complete series is then defined and we have a 'route of approximation' [PNK, p. 76].

The analogy drawn here can be most illuminating provided it is not pressed too closely. In the theory of limits one is concerned with real numbers which, in a certain well-defined sense, are "limits" of certain well-defined infinite sequences of real numbers. Both the numbers themselves and the infinite sequences of numbers are abstract entities susceptible of extremely precise definition (although both concepts were successfully used by mathematicians long before the precise definitions had been formulated). In the method of extensive abstraction, on the other hand, one is concerned with *perceived* events and with indefinitely prolonged sequences of these events (abstractive classes of events). Neither the events (which, it will be recalled, are the only fully actual natural entities, according to Whitehead) nor the indefinitely prolonged sequences of events are susceptible of precise definition because of the inherent limitations of sense-perception. However, *some* of the events in an abstractive class are perceived through the senses as extending over one another; and the general concepts of events and of extension between events are clear enough to make the search for convergence to simplicity an "instinctive procedure of habitual experience." This instinctive procedure is lifted to the level of reflective awareness by the method of extensive abstraction without, however, any serious loss of contact with immediate sense-experience. But this very contact with sense-experience precludes the type of precision possible when dealing with pure mathematical entities such as numbers. Abstractive classes, in a suggestive phrase of Whitehead's, "guide thought" (CN, p. 61)—guide thought, I would add, in directions determined by sense-perception but to destinations beyond all possibility of detection by sense-perception. These destinations are, of course, the abstractions of science, "entities which are truly in nature, though they have no meaning in isolation from nature" (CN, p. 173).[38]

It often seems difficult to reconcile Whitehead's insistence on direct perception as the source of scientific concepts with his complex logical and mathematical devices for constructing and elaborating these concepts.[39] One of the most extreme instances of this difficulty occurs in the case of puncts, rects, and levels. It certainly seems amazing at first sight to find the points, lines, and planes of theoretical physics explicated in terms of intersections of moments of different time-systems—after all, since an observer can be in only *one* time-system at a time, the perception of the intersections of moments of several time-systems would seem to be impossible in principle; and furthermore, points, straight lines, and planes were used successfully in science long before anyone even dreamed of the possibility of multiple time-systems in nature. Are we forced then to conclude that in this case Whitehead's preoccupation with ingenious mathematical constructions led him to forget his own epistemological dicta? Evidence pointing to an affirmative answer to this question might be provided by the fact that Whitehead seems eventually to have abandoned his definitions of puncts, rects, and levels. It will be recalled that the principal use of puncts, rects, and levels was to define the "points," "straight lines," and "planes" of space-time and then points, straight lines, and planes (in time-less space). Now, in Whitehead's later treatment of the method of extensive ab-

[38] Cf. Whitehead's characterization in PR (p. 496) of the extensive relationships in terms of which he there defines straight lines: "These relationships are *there* for perception." Here I interpret Whitehead to mean that the extensive relationships in question are systematically illustrated in sense-perception in such a way as to "guide thought" to the concept of a straight line.

[39] Whitehead defends his definition of points from the charge of paradox in the following terms: "If there is no absolute position, a point must cease to be a simple entity. What is a point to one man in a balloon with his eyes fixed on an instrument is a track of points to an observer on the earth who is watching the balloon through a telescope, and is another track of points to an observer in the sun who is watching the balloon through some instrument suited to such a being. Accordingly if I am reproached with the paradox of my theory of points as classes of event-particles, and of my theory of event-particles as groups of abstractive sets, I ask my critic to explain exactly what he means by a point. While you explain your meaning about anything, however simple, it is always apt to look subtle and fine spun. I have at least explained exactly what I do mean by a point, what relations it involves and what entities are the relata. If you admit the relativity of space, you also must admit that points are complex entities, logical constructs involving other entities and their relations" (CN, pp. 135–36).

straction in PR, the replacement of extension by the new concept of "extensive connection" (a binary relation with formal properties slightly different from extension) appears to make possible the definition of "points," "straight lines," and "planes" of space-time independently of puncts, rects, and levels. One might, therefore, be tempted to conclude that Whitehead decided to eliminate the objectionable definitions which construed puncts, rects, and levels as intersections of non-parallel moments. A further conclusion would be that perhaps Whitehead no longer requires a multiplicity of time-systems for the definition of fundamental geometrical entities, which would in turn imply that spatial order is perhaps independent of temporal order. Before accepting the idea that Whitehead really intended such sweeping revisions in his method of extensive abstraction, let us examine the whole matter more carefully.

In the first place Whitehead's explicitly stated view of the epistemological status of levels, rects, and puncts must not be ignored:

> Evidently levels, rects, and puncts in their capacity as infinite aggregates cannot be the termini of sense-awareness, nor can they be limits which are approximated to in sense-awareness. Any one member of a level has a certain quality arising from its character as also belonging to a certain set of moments, but the level as a whole is a mere logical notion without any route of approximation along entities posited in sense-awareness [CN, p. 92].

There can be no doubt, then, that Whitehead was not deceiving himself about the observability of intersections of non-parallel moments. The observable basis for points, straight lines, and planes lies rather in event-particles (including certain special types of loci of event-particles). That event-particles (i.e., "points" in space-time) can be defined independently of puncts is undoubtedly an advance in the method of extensive abstraction, but it by no means makes puncts dispensable. To clarify this point we must refer again to the three characters of any geometrical entity, namely, "(i) its extrinsic character which is its character as a definite route of convergence among events, (ii) its intrinsic character which is the peculiar quality of nature in its neighbourhood, namely, the character of the physical field in the neighbourhood, and (iii) its position" (CN, p. 191). Thus, the *complete* meaning of a point (as it occurs in theoretical physics) involves not only *what* a point is (extrinsically in terms of a set of abstractive classes and intrinsically in terms of the physical properties and relations which are the limits of the physical properties and relations possessed by the members of any one of these abstractive classes) but also *where* it is in the instantaneous space of some time-system. And Whitehead contends that the meaning of the position of a point in space can only be conceived as the intersection of a group of (four) non-parallel moments. If the positional character of event-particles must derive from puncts, one might surmise that the "points" of PR are entirely lacking in positional character. This is in fact the case, as we shall see in chapter vi. The superiority of the later concept of points over the earlier concept of event-particles lies not in the elimination of puncts (the source of spatial position or spatial *order*), but rather in the sharper separation between the several distinct and equally real aspects of the ultimately simple geometrical entities. The new definitions of point, straight line, and plane in PR ignore the metrical as well as the ordinal properties of these entities and concentrate upon their "qualitative" (or more precisely, topological and projective) properties. The latter are perhaps identical with what were above termed the "extrinsic" aspects of a geometrical entity; the "intrinsic" aspects presumably include both the metrical properties of space-time and the properties of the physical field. I would suggest (and here I anticipate the discussions of chapter vi) that Whitehead's new definitions of points, straight lines, and planes are not so much designed to replace the older definitions as to serve a different purpose in a new, metaphysically oriented context.

Assuming, then, that levels, rects, and puncts are not dispensable in deriving the properties of space, time, and motion, we revert to our earlier question: how, we wonder, was the position of a point conceived in pre-relativity physics before the possibility of multiple time-systems in nature had even been envisaged? Historically, two different answers were given to this question. Space might be conceived as absolute, so that each point in one of the relative spaces actually employed by physicists obtained its positional character ultimately from its location in the single eternal absolute space. If absolute space were rejected, the points of each relative space might be thought of as specified by means of their relations to bodies (i.e., persisting bits of matter), most fundamentally by their relations to a "reference-body" at the origin of some "coordinate-system." We have already discussed Whitehead's objections to each of these two traditional theories of space (see above p. 6), though of course it is to be understood that despite their ultimate failure these conceptions of space were, up to a point, highly useful in the development of science. We are forced, then, to develop a theory of space in which the relata are events, and the only way of successfully doing this is to construct a whole family of relative spaces simultaneously by the method of intersecting moments.[40] Such construction is epistemologically justified as soon as one recognizes the possibility of alternative time-systems, each correlated with a definite state of motion of the observer: since alternative states of motion are readily conceived, so are alternative time-systems and hence the notion of intersecting moments. Whitehead's theory of space possesses, furthermore, the great philosophical virtue of "coherence" (where "incoherence" means "the arbitrary disconnection of first principles" [PR, p. 9]). Thus Whitehead says:

. . . by the aid of the assumption of alternative time-systems, we are arriving at an explanation of the character of space. In natural science 'to explain' means merely to discover 'interconnexions.' . . . the dependence of the character of space on the

[40] It is no longer open to us to begin by "identifying" an event as origin and then to go on to "locate" all the rest of our points by means of length-measurements (with, say, rigid rods), since an *event* does not persist through time. At best one might hope in this way to obtain an instantaneous space which, however, could not be related to any other instantaneous space; and if one assumes that measurement is a process which takes time, one could not in this way obtain even an instantaneous space.

character of time constitutes an explanation in the sense in which science seeks to explain. The systematising intellect abhors bare facts. The character of space has hitherto been presented as a collection of bare facts, ultimate and disconnected. The theory which I am expounding sweeps away this disconnexion of the facts of space [CN, pp. 97–98].

I have been concerned in the preceding discussion to forestall one type of misunderstanding of Whitehead's procedure in deriving the basic results of space-time geometry (including kinematics) by the method of extensive abstraction. There remains, however, as I see it, an important unresolved dilemma in that procedure. I have repeatedly stressed the fact that Whitehead is interested in developing a general approach to space-time geometry which has its basis in a sound natural philosophy— an approach which will, therefore, be compatible with the metaphysical principles of the philosophy of organism. But Whitehead *also* claims that the particular form which his general approach takes—the actual procedure followed in applying the method of extensive abstraction— provides an explanation of space and time as used by physicists and of the results of the Michelson-Morley and other ether-drift experiments:

. . . if we allow this possibility [of different time-systems] we not only explain many modern delicate experiments, but we also obtain explanations of what we mean by the spatial extension in three dimensions, and by planes and straight lines, and parallels and right-angles [R, p. 59].[41]

The dilemma to which I have referred may, then, be put as follows: either Whitehead's procedure depends upon the assumption that there really are alternative time-systems in nature (in which case he should show how direct sense-perception furnishes evidence for this assumption); or Whitehead's procedure depends only on the assumption of the mere abstract possibility of alternative time-systems in nature (in which case his procedure is incapable of singling out hyperbolic kinematics as the real kinematics of nature). But Whitehead admits that direct sense-perception furnishes no evidence for the actual existence of alternative time-systems in nature and yet he wishes to account for the actual concepts of space, time, and motion, and some of the actual experimental results, of contemporary physics. Another way of making my point is to ask if Whitehead's procedure could explain parabolic (i.e., classical Galilean) kinematics, supposing all the experimental evidence were in its favor? Surely, one could not logically explain kinematic formulae which imply just a single time-system in nature by means of a procedure which presupposes—even as a mere abstract possibility—a multiplicity of time-systems.

The two possible ways of resolving this dilemma are, I believe, both unsatisfactory (but perhaps not equally so) from the point of view of Whitehead's natural philosophy. We might, on the one hand, accept the multiplicity of time-systems as a mere possibility; but then Whitehead's entire procedure seems to

---

[41] Cf. CN (p. 192), where Whitehead refers to "the sources [including multiple time-systems] from which the whole of geometry receives its physical explanation."

become little else than a more or less impressive logico-mathematical technique without any clear-cut relevance to the concepts and theories of contemporary physics. On the other hand, we might take the existence of multiple time-systems in nature as a fundamental principle of natural philosophy (whether warranted by direct sense-perception or not). This latter alternative seems clearly incompatible with such remarks as the following: "In an organic philosophy of nature there is nothing to decide between the old hypothesis of the uniqueness of the time discrimination and the new hypothesis of its multiplicity. It is purely a matter for evidence drawn from observations" (SMW, pp. 176–77). (Here "observations" can only refer to the highly indirect and inferential experimental methods of modern physics, since "observation" in the form of direct sense-perception is basic to Whitehead's organic philosophy of nature.) Again, Whitehead says that the "metaphysical hypothesis" of multiple time-systems is introduced "in order to satisfy the present demands of scientific hypothesis" (SMW, p. 181)—which certainly suggests that hyperbolic kinematics (the scientific hypothesis in question) is *not* a fundamental datum of sense-perception.

The ultimate and irreducible in Whitehead's analysis of kinematics are events and their properties (the constants of externality); thus, when Whitehead, in the summary passage about to be quoted, attributes the possibility of multiple space-time systems to the existence of event-particles, we realize that the *ultimate* source of this possibility lies in the existence of the events in terms of which event-particles are defined.

. . . event-particles are the ultimate elements of the four-dimensional space-time manifold which the theory of relativity presupposes. You will have observed that each event-particle is as much an instant of time as it is a point of space. I have called it an instantaneous point-flash. Thus in the structure of this space-time manifold space is not finally discriminated from time, and the possibility remains open for diverse modes of discrimination according to the diverse circumstances of observers. It is this possibility which makes the fundamental distinction between the new way of conceiving the universe and the old way. The secret of understanding relativity is to understand this [CN, p. 173].

Our dilemma remains unresolved. Whatever be one's final judgment on the philosophical import of the method of extensive abstraction, it is clear that one is dealing here with a critical issue in almost the entire tradition of Western philosophy, namely, the question of the appropriate kind and degree of influence which "science" should exert on "philosophy." Like many other philosophical schemes in this tradition, Whitehead's philosophy of science deliberately attempts at once to transcend the particularities of current science and somehow to illuminate or make intelligible those very particularities.

CHAPTER VI

THE METHOD

OF EXTENSIVE

ABSTRACTION

AND THE

PHILOSOPHY

OF ORGANISM

1. Extensive Connection and Regions; the Extensive Continuum

In PR (Part IV, "The Theory of Extension"), Whitehead modifies his method of extensive abstraction in the direction of greater generality (in fact, as we shall see, there seems good reason to believe that the new formulation is entirely too general to achieve Whitehead's objectives). Instead of postulating a four-dimensional pseudo-Euclidean structure for space-time (as in the formulations described in chapter v), Whitehead succeeds in defining the non-ordinal and non-metrical properties of points, straight lines, and planes on the basis of the single fundamental idea of the possible "connections" among "regions." In the manifold of regions space and time are not initially differentiated; and no special types of regions (such as durations) need be presupposed. In order to elucidate the significance of the formal developments I shall begin by characterizing the two undefined concepts, "extensive connection"[1] and "region."[2]

Extensive connection is a binary relation whose relata are (bounded and closed)[3] *regions*. Two regions are extensively connected when, roughly speaking,

[1] This concept was first introduced by T. de Laguna (from whom Whitehead borrows it— see PR, p. 440) in "Point, Line, and Surface, as Sets of Solids."

[2] Later (see below, pp. 145–46) we shall see that regions may be *defined* in terms of the categories of the philosophy of organism. This definition, though metaphysically significant, is irrelevant to the method of extensive abstraction.

[3] Cf. PR, p. 546: ". . . a certain determinate boundedness is required for the notion of a region. . . . The inside of a region, its volume, has a complete boundedness denied to the extensive potentiality external to it. The boundedness applies both to the spatial and the tem-

they have at least one point in common. (It must be emphasized that the last assertion is made solely in order to aid our intuitive understanding of extensive connection; such assertions have no part at all in the formal development of the subject.) Thus any two regions which either wholly or partly overlap, or are in contact along parts of their boundaries or at a single point, are extensively connected (Figure 18). Before stating some of the formal assumptions[4] about extensive connection (which we call, for short, simply *connection*), it is con-

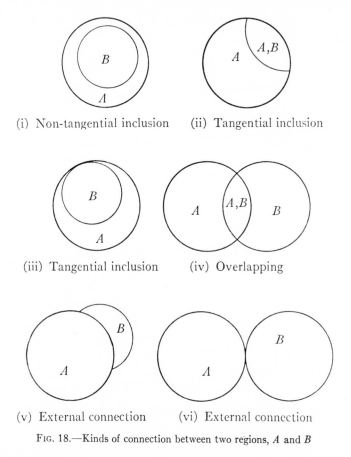

(i) Non-tangential inclusion   (ii) Tangential inclusion

(iii) Tangential inclusion   (iv) Overlapping

(v) External connection   (vi) External connection

Fig. 18.—Kinds of connection between two regions, *A* and *B*

poral aspects of extension. Wherever there is ambiguity as to the contrast of boundedness between inside and outside, there is no proper region." In short, a "bounded" region is one that possesses some definite "boundary." A "closed" region is a bounded region that includes its boundary. Whitehead never says explicitly that his regions are closed, but it is a reasonable inference from the properties he does attribute to regions. In terms of standard mathematical conceptions, regions seem to be purely topological in character; this question, however, will receive extended discussion later on (see below, pp. 110–11).

[4] Whitehead does not bother to differentiate axioms from theorems (with one exception) in his formulations in PR (Part IV, chaps. ii and iii).

venient to define another closely related concept. Two regions are *mediately connected* when they are both connected with some third region (Figure 19). We assume that both connection and mediate connection are symmetrical relations; that no region is connected with all other regions but that any two regions are mediately connected; that connection is not transitive (but not intransitive either); that no region is connected or mediately connected with itself. We may now define a relation which closely resembles the "extension" (or K) of Whitehead's earlier works: Region A *includes* region B (or B is a *part of A*) when every region connected with B is also connected with A, but not every region connected with A is also connected with B.[5] Further assumptions about inclusion are: when one region includes another, the two regions are connected; inclusion is transitive, irreflexive, and asymmetrical; every region includes other regions, among them pairs of non-connected regions (Figure 18, i–iii).

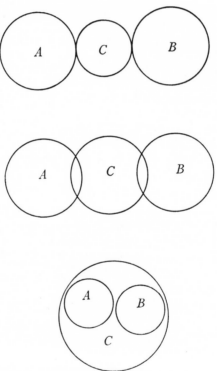

FIG. 19.—Kinds of mediate connection between two regions, A and B

[5] This is Whitehead's Definition 2 (PR, p. 452) except that he omits the second clause—clearly an oversight, since Assumption 7 (PR, p. 452) asserts that a region does *not* include itself, contrary to Definition 2. Furthermore, de Laguna's definition of the same concept (which Whitehead says he has adopted) corresponds to that in the text above (see de Laguna, "Point, Line, and Surface, as Sets of Solids," Definition III, p. 452).

Before proceeding with further assumptions and definitions, I shall discuss the nature of regions and of extensive connection in order to clarify the new philosophical context in which the method of extensive abstraction is now being developed. First of all, the formal properties of regions may be compared with those of events in the earlier works. It seems clear that Whitehead intends regions (the relata of extensive connection) to be formally almost identical with events (the relata of extension). It is impossible to demonstrate this formal identity between regions and events because Whitehead never lays down a complete set of axioms for either concept. However, the following considerations should be sufficiently convincing. We begin by assuming that extension and inclusion are essentially synonymous (both Whitehead and de Laguna, from whom he borrowed the definition of inclusion, tell us this). We then find that the first five of our eight axioms for extension (see p. 44) have the following close counterparts for inclusion.

Axiom i—Definition 2 (PR, p. 452).
Axiom ii—Assumption 9 (PR, p. 452).
Axiom iii—Assumption 6 (PR, p. 452).
Axiom iv—Non-abstractive Condition viii (PR, p. 463).
Axiom v—Non-abstractive Condition vii (PR, p. 463).

Assumption 9 actually corresponds to just the first clause in Axiom ii, but an assertion corresponding to the second clause of Axiom ii can easily be deduced from Non-abstractive Condition vii (simply let the finite number of regions referred to in Non-abstractive Condition vii reduce to one). As for the remaining three axioms for extension, the counterpart of the sixth is surely meant to hold for extensive connection even though it is nowhere explicitly stated; while it is pretty clear from Whitehead's diagrams illustrating inclusion, overlapping, and external connection (reproduced in Figure 18) that counterparts of Axioms vii and viii (concerning junction and injunction) must hold for regions.[6] Finally, there do not seem to be any important assumptions about regions which are not also true of events.

One further point must be considered: some of the assumptions concerning inclusion in the above list are not formulated so as to apply to *all* regions. Thus, while Assumption 6 says that the relation of inclusion is transitive and Assumption 9 says that every region includes other regions, Non-abstractive Conditions vii and viii refer only to members of a special class of regions in the extensive continuum (a so-called "ovate" class of regions, which will be defined below). Whitehead's procedure here is in accord with the broad metaphysical outlook of PR: the assumptions which apply to all regions are the ones required to define an abstractive class of regions—the basic concept of the method of extensive abstraction; the assumptions which apply only to the regions in an ovate class are the additional ones required to define certain specific geometrical

[6] Upon comparing Figure 18 with Figure 6, one will find that tangential inclusion (ii) corresponds to injunction; that overlapping (iv) corresponds to junction of intersecting events; and that external connection (v) corresponds to adjunction.

entities (e.g., straight lines) which may possibly be exemplified only in certain cosmic epochs—hence Whitehead leaves open the question of whether these assumptions really apply to all regions. It turns out, then, that the sole formal difference between regions and events which is explicitly mentioned by Whitehead is the fact that regions are limited in extent or bounded, whereas events may be (as in the case of durations) unbounded. We turn next to the non-formal aspects of regions.

Regions are in, or of, the *extensive continuum*, which refers to the complex scheme of extensional (i.e., spatial and temporal—but the two have yet to be distinguished) relations, such as whole-part, overlapping, and contact, which serves to unite all real entities. The extensive continuum is characterized in some detail by the following passage:

... the real potentialities relative to all standpoints are coordinated as diverse determinations of one extensive continuum. This extensive continuum is one relational complex in which all potential objectifications find their niche. It underlies the whole world, past, present, and future. Considered in its full generality, apart from the additional conditions proper only to the cosmic epoch of electrons, protons, molecules, and star-systems, the properties of this continuum are very few and do not include the relationships of metrical geometry. An extensive continuum is a complex of entities united by the various allied relationships of whole to part, and of overlapping so as to possess common parts, and of contact, and of other relationships derived from these primary relationships. The notion of a 'continuum' involves both the property of indefinite divisibility and the property of unbounded extension. There are always entities beyond entities, because nonentity is no boundary [PR, p. 103].

The most general characteristics of the extensive continuum—those character-istics to which we can conceive no alternatives—are "ultimate metaphysical necessities" (PR, p. 441); these include indefinite divisibility and unbounded extension (either finite or infinite) but not shape or dimensionality or straight-ness or measurability; in our own particular cosmic epoch, on the other hand, the extensive continuum is a four-dimensional metrical manifold with three spatial dimensions and one temporal dimension.[7]

At this point it begins to look very much as if Whitehead's notion of a region is purely topological, i.e., regions are assumed to possess just those properties which are invariant with respect to topological transformations. Now we have

---

[7] It is perhaps not easy to conceive of a real, physical manifold without dimensionality or even with different dimensions in different parts. If Whitehead is wrong about the "conceiv-ability" of such notions, that simply means that a fixed number of dimensions is an ultimate metaphysical necessity of the universe. Cf. MT, p. 79: "... this planet, or this nebula in which our sun is placed, may be gradually advancing towards a change in the general character of its spatial relations. Perhaps in the dim future mankind, if it then exists, will look back to the queer, contracted three-dimensional universe from which the nobler, wider existence has emerged.

"These speculations are, at present, neither proved or disproved. They have however a mythical value. They do represent how concentration on coherent verbalizations of certain aspects of human experience may block the advance of understanding."

already encountered metrical and affine transformations in the previous chapter, and later on in this chapter we shall encounter the even more general class of projective transformations. Topological transformations, however, are far more general than any of these: roughly speaking, any arbitrary deformation of a geometrical entity (e.g., a sphere) which does not "tear" it and which does not bring distinct points into coincidence is a topological transformation (but there are topological transformations which are not deformations). The precise analytical definition is as follows: A topological point transformation is one-one and continuous. (Metrical, affine, and projective transformations are all special cases of topological transformations.) Examples of topologically equivalent figures are a circle, an ellipse, and a triangle; examples of topologically non-equivalent surfaces are the respective surfaces of a sphere and a torus. (In Figures 18 and 19 I have reproduced Whitehead's diagrams [PR, pp. 450–51] but it is essential to realize that if I am right about the topological character of Whitehead's regions, then a circle in these diagrams must be understood to stand for *any other topologically equivalent figure*—which means, any figure into which a circle can be continuously distorted, e.g., an ellipse or an hour-glass shape.)

To complete my non-formal characterization of regions, I shall attempt to place the concept of the extensive continuum in its metaphysical context. First of all, the extensive continuum should be compared with the still more general concepts of *creativity* and *the primordial nature of God*. " 'Creativity' is the universal of universals characterizing ultimate matter of fact" (PR, p. 31). Creativity is common to God and creatures and probably may be identified with what Whitehead refers to as "the one underlying activity of realisation individualising itself in an interlocked plurality of modes" (SMW, p. 102). Thus, creativity is not only a necessary feature of any possible world but it is the single trait common to *all* actualities. The primordial nature of God is usually treated by Whitehead as the complement of creativity. Unlimited creativity would produce a chaotic world—or, rather, no world at all, for there would be no positive force to choose among contrary alternatives. "From this point of view," says Whitehead, "he [God] is the principle of concretion—the principle whereby there is initiated a definite outcome from a situation otherwise riddled with ambiguity" (PR, p. 523). The primordial nature of God involves the realm of eternal objects (or pure potentials):[8] "Viewed as primordial, he is the unlimited conceptual realization of the absolute wealth of potentiality. In this aspect, he is not *before* all creation, but *with* all creation" (PR, p. 521). And, "by reason of this primordial actuality, there is an order in the relevance of eternal

---

[8] Eternal objects are, ontologically, the polar opposites of actual occasions (on which see the next note), and may be thought of as "forms of definiteness" or "universals" (PR, pp. 32, 76). There are two principal types of eternal objects, those of the "objective (or mathematical) species" (e.g., geometric entities and relationships) and those of the "subjective species" (e.g., "an emotion, or an intensity, or an adversion, or an aversion, or a pleasure, or a pain" [PR, p. 446]).

objects to the process of creation. . . . The primordial nature of God is the acquirement by creativity of a primordial character" (PR, p. 522).

The mere existence of any possible world depends upon creativity and God's primordial nature, whereas the most general "objective" or "mathematical" features of such a world are expressed by the extensive continuum. "This extensive continuum expresses the solidarity of all possible standpoints throughout the whole process of the world. It is not a fact prior to the world; it is the first determination of order—that is, of real potentiality—arising out of the general character of the world" (PR, p. 103). The extensive continuum in itself is merely divisible; this potentiality for division is actualized by the concrescence of actual entities, which thereby atomize the continuum. One may ask, how is this merely *potential* continuum *real*, and the answer is, "because it expresses a fact derived from the actual world and concerning the contemporary actual world. All actual entities are related according to the determinations of this continuum; and all possible actual entities in the future must exemplify these determinations in their relations with the already actual world" (PR, p. 103). Thus the relations of the extensive continuum are not *pure* potentials but rather real potentialities arising directly out of the immediately preceding actual world but indirectly out of the entire past world, and referring indirectly to the entire future world.

The extensive continuum is a natural outgrowth of Whitehead's attempt to reconcile atomism and continuity in nature. It has already been mentioned that in his early writings there is no doctrine of minimum events; in fact, the method of extensive abstraction there explicitly repudiates the idea of either minimum or maximum events. Also, Whitehead then believed that his axioms for extension involved an expression of the continuity of nature—what he called an "ether of events." The adoption of an "epochal theory of time" and of a corresponding "cell theory of actual occasions"[9] requires a different way of explaining the continuity of nature. This explanation involves assigning the concept of potentiality a more fundamental place in Whitehead's philosophy than it ever occupied before. The extensive continuum is an expression of this new emphasis on potentiality as the essence of continuity.[10] Thus a region is said

---

[9] Cf. PR, p. 53: ". . . the ultimate metaphysical truth is atomism. The creatures are atomic. In the present cosmic epoch there is a creation of continuity. Perhaps such creation is an ultimate metaphysical truth holding of all cosmic epochs; but this does not seem to be a necessary conclusion. The more likely opinion is that extensive continuity is a special condition arising from the society of creatures which constitute our immediate epoch." The "creatures" in this passage are *actual occasions*, i.e., the ultimate realities of the universe. Whitehead characterizes them as follows: " 'Actual entities'—also termed 'actual occasions'—are the final real things of which the world is made up. There is no going behind actual entities to find anything more real. They differ among themselves: God is an actual entity, and so is the most trivial puff of existence in far-off empty space" (PR, pp. 27–28).

[10] Cf. PR, p. 96: "These possibilities of division [of the contemporary world] constitute the external world a continuum. For a continuum is divisible; so far as the contemporary world is divided by actual entities, it is not a continuum, but is atomic. Thus the contemporary world is perceived with its potentiality for extensive division, and not in its actual atomic division."

It will be noted that we have here very nearly a reversal of Whitehead's earlier view which

to be "an extensive standpoint in the real potentiality for actualization" (PR, p. 546). From the ontological point of view of Whitehead's philosophy of organism, an "extensive standpoint" must be understood as one of the multiplicity of possible ways in which an actual occasion might *prehend* (or *feel*)[11] the extensive continuum as a whole; or, more loosely, one of the possible spatio-temporal perspectives from which an observer might view the rest of the world. A prehension of the sort just mentioned would have as its data a complex variety of extensive relationships (eternal objects of the mathematical species), the most vivid of which would normally be those exemplified in its immediate spatio-temporal environment. However, what has just been said applies only to an *actual* region, i.e., to an *actual* standpoint in the extensive continuum of some definite actual occasion. *Potential* standpoints in the extensive continuum, and thereby *potential* regions, must be accounted for—and without violating one of Whitehead's fundamental metaphysical principles, that the potential must always be somehow relative to the actual.[12] Whitehead appeals here to another of his "categories of existence" (we have already encountered actual occasions, eternal objects, and prehensions), namely, "contrasts." A *contrast* is a mode of synthesis of diverse types of entities in one prehension. In the context we are considering, Whitehead introduces the concept of a *coordinate division*, which is (roughly) a contrast with two components: (1) an actual occasion occupying a definite region called the *basic region*, and (2) the potentiality of that actual occasion having arisen from the extensive standpoint of some subregion of the basic region. It can be seen that a coordinate division is an extremely complicated entity: it involves both the physical prehension of an actual occasion and the "impure" (physical-mental) prehension of a real potentiality. Briefly, coordinate divisions are *feelings* whose data are the potential or hypothetical subdivisions of the region occupied by an actual occasion—" 'feelings which might be separate' " (PR, p. 436).

In order to obtain the complete scheme of extensional relationships which characterizes the extensive continuum, provision must be made for aggregation as well as for subdivision of actual occasions. Hence Whitehead's remark that "in addition to the merely potential subdivisions of a satisfaction into coordinate feelings, there is the merely potential aggregation of actual entities

---

locates the sources of continuity and of actuality in events and the sources of atomicity and of possibility in objects (see above, pp. 23–24). It is impossible to examine here the metaphysical significance of this reversal.

[11] ". . . every prehension consists of three factors: (a) the 'subject' which is prehending, namely, the actual entity in which that prehension is a concrete element; (b) the 'datum' which is prehended; (c) the 'subjective form' which is *how* that subject prehends that datum" (PR, p. 35). A *feeling* is a *positive* prehension, i.e., a prehension whose datum is incorporated in the progressive concrescence of the subject; in the case of a *negative* prehension, its datum is not so incorporated. Cutting across the distinction between positive and negative prehensions is the distinction between "physical" and "conceptual" prehensions. A *physical prehension* has as its datum an actual occasion; a *conceptual prehension* has as its datum an eternal object.

[12] This is the "ontological principle," which states that "actual entities are the only *reasons;* so that to search for a *reason* is to search for one or more actual entities" (PR, p. 37).

into a super-actuality in respect to which the true actualities play the part of coordinate subdivisions" (PR, p. 439). As described by Whitehead, both sub-division and aggregation would seem to presuppose *past* actual occasions as a source of data for the present coordinate feelings—which leaves him with the problem of explaining the reality of present and future extensional relations in the extensive continuum. In somewhat more general terms, the problem can be put like this: the fundamental type of relation responsible for connecting and unifying the various parts of the universe is the influence (called *causal ob-jectification*) of earlier on later actual occasions.[13] But, first of all, since (in ac-cordance with the views of modern physics) there is assumed to be an upper limit to the speed of transmission of such influence, the solidarity of the present state of the universe is not accounted for; and second, the reality of the future seems utterly without foundation. From the point of view of the natural sci-ences, there is a special urgency about the first of these difficulties because measurement always takes place in the space defined by the present state of the universe; while from the metaphysical standpoint of the philosophy of organ-ism, there is a special urgency about the second of these difficulties because the ontological status of the future obviously has profound implications for the meaning of such concepts as creativity, freedom, and determinism. White-head's detailed resolution of these two difficulties cannot be discussed here. It is worth remarking, however, that he goes so far as to introduce a special type of objectification, derivative from the primary causal type, and concerned with the objectification of contemporary actual occasions. If such *presentational ob-jectification* (PR, p. 91) becomes conscious, one has perception in the mode of *presentational immediacy* (on which see below, p. 140), and by means of such perception there is exhibited "the community of contemporary actualities as a common world with mathematical relations . . ." (PR, p. 97). As for the status of the future, Whitehead deals with it largely in terms of the *anticipation* by present actual occasions of certain general features of future actual occasions. In particular, there is anticipation of the general extensional relations of the future, and thereby in a certain sense present extensional relations are "extrap-olated" into the future. The following statement concerning the ontological status of the future could, I believe, be readily recast so as to apply to the special case of future *extensional* relations.

. . . there are no actual occasions in the future, already constituted. Thus there are no actual occasions in the future to exercise efficient causation in the present. What is

[13] "The term 'objectification' refers to the particular mode in which the potentiality of one actual entity is realized in another actual entity" (PR, p. 34). Objectification is the converse of prehension: earlier actual entities are objectified in later ones; later actual entities prehend earlier ones.

The extensive continuum is "the primary factor in objectification" in the sense that "It provides the general scheme of extensive perspective which is exhibited in all the mutual objectifications by which actual entities prehend each other. Thus, in itself, the extensive continuum is a scheme of real potentiality which must find exemplification in the mutual prehension of all actual entities" (PR, p. 118).

objective in the present is the necessity of a future of actual occasions, and the necessity that these future occasions conform to the conditions inherent in the essence of the present occasion. The future belongs to the essence of present fact, and has no actuality other than the actuality of present fact. But its particular relationships to present fact are already realized in the nature of present fact [AI, p. 251].

We see from the above discussion that some regions are "potential" or "hypothetical" entities, unlike events which, it will be recalled, are all equally "concrete" or "actual" entities. This shift in the nature of the fundamental entities dealt with by extensive abstraction has its roots in the following principle of Whitehead's philosophy of organism: the "internal" and unextensive aspect of an actual entity is an indivisible unity; the "external" and extensive aspect of an actual entity is indefinitely divisible.[14] Thus, in his early analysis of natural science, events are taken as the most fully actual entities *in nature*[15] (the "internal" aspect of reality being deliberately ignored), and events in this sense are indefinitely divisible into other events; while in his later metaphysics, individual occasions of experience are taken as the ultimately actual entities (combining both the "internal" and the "external" aspects of reality) and such occasions are in no sense divisible into other occasions, although their respective regions are indefinitely divisible into other regions. Now, one of the fundamental demands of the method of extensive abstraction is that its extended elements or entities be indefinitely divisible; we have just seen that this demand can only be satisfied by the substitution of regions for events when the ontological categories implicit in Whitehead's natural philosophy become the explicit ontological categories of his metaphysics.

Can we also explain the exclusion of unbounded regions in ontological terms? To say that all regions are bounded is to say that each actual occasion must

---

[14] The common-sense way of putting Whitehead's principle would be to say that the successively smaller parts of an organism are not generally themselves organisms whereas the successively smaller parts of a non-organic whole are—up to a point, at least—homogeneous with the whole itself. Whitehead's technical language is this: ". . . it is only the physical pole of the actual entity which is thus divisible. The mental pole is incurably one . . . . for many abstractions concerning low-grade actual entities, the coordinate divisions approach the character of being actual entities on the same level as the actual entity from which they are derived.

"It is thus an empirical question to decide in relation to special topics, whether the distinction between a coordinate division and a true actual entity is, or is not, relevant. In so far as it is not relevant we are dealing with an indefinitely subdivisible extensive universe" (PR, pp. 436–37).

[15] There are numerous statements to this effect in Whitehead's early writings on philosophy of science; perhaps the most explicit is the following: ". . . our lowest, most concrete, type of abstractions whereby we express the diversification of fact must be regarded as 'events,' meaning thereby a partial factor of fact which retains process" (PA, p. 223). This suggests a possible relationship between the events of Whitehead's early writings and the actual occasions of his philosophy of organism, namely, an event is an aspect of an actual occasion. More specifically, an event is, or is closely related to, the physical pole (see previous note) of an actual occasion. Supporting this suggestion is the fact that both events and physical poles are indefinitely divisible.

represent a definite and limited standpoint in the extensive continuum, and this does seem to be in accordance with Whitehead's conception of actual occasions. However, actual occasions do not exhaust the category of actual entities in Whitehead's metaphysical scheme; there is also the "primordial" actual entity, God (PR, pp. 134–35). But God does not appear to be extended in the sense of occupying a region; for God's region would presumably have to be identified with the entire extensive continuum, and it is hard to imagine that Whitehead would be (as he is) silent about such an important identification. Furthermore, even indefinitely repeated aggregation of actual occasions—always, of course, within a single prehension of a single actual occasion—could never lead to an unlimited region (a possible exception in the case of elliptic, and therefore finite, space-time would be associated at best only with certain unusual actual occasions). Thus, there seem to be compelling reasons for excluding unbounded regions (analogous to the durations of the early works) from the list of fundamental ontological categories of the philosophy of organism. This exclusion has consequences for the new formulation of extensive abstraction: in order to define points without durations, Whitehead replaces the relation of extension by the relation of extensive connection. We shall now examine the significance of this replacement.

In the earlier development of extensive abstraction "subdivision" and "aggregation" were not distinguished—given the character of events there was not even the possibility of such a distinction—but the former operation was subtly emphasized at the expense of the latter by the choice of extension (essentially a whole-part relation) as a primitive concept. On the other hand, the primitive concept, extensive connection, subtly emphasizes connections between wholes rather than parts of wholes. I cannot see, however, any sense in which the characteristic philosophical doctrines of the early works imply a preference for "subdivision" over "aggregation." Nor is the concept of extensive connection in any way implied by the characteristic doctrines of PR. It is true that from the metaphysical point of view which pervades PR extensive connection is a particularly happy invention, since it permits a marked gain in generality insofar as "the somewhat more general notion of 'extensive connection' can be adopted as the starting point for the investigation of extension; and . . . the more limited notion of 'whole and part' can be defined in terms of it" (PR, p. 440).[16] Furthermore, once the notion of whole and part is available, points and the other basic geometrical entities of space-time may be defined independently of moments and therefore of durations. In the following passage Whitehead contrasts the old and new methods of extensive abstraction.

If we confine our attention to the subdivision of an actual entity into coordinate parts, we shall conceive of extensiveness as purely derived from the notion of 'whole and part,' that is to say, 'extensive whole and extensive part.' This was the view taken in

---

[16] Thus Whitehead accepts de Laguna's judgment that "It is . . . impossible by means of the method in its original form to give a definition of the point in terms of the solid" ("Point, Line, and Surface, as Sets of Solids," p. 451).

by [*sic*] my two earlier investigations of the subject [PNK and CN]. This defect of starting point revenged itself in the fact that the 'method of extensive abstraction' developed in those works was unable to define a 'point,' without the intervention of the theory of 'duration.' Thus what should have been a *property* of 'durations' became the definition of a point. By this mode of approach the extensive relations of actual entities mutually external to each other were pushed into the background; though they are equally fundamental [PR, pp. 439–40].

Some care is necessary in interpreting this passage. Whitehead is not, I believe, repudiating the concepts defined in the earlier formulation of extensive abstraction: levels, rects, and puncts (or some equivalent set of concepts) still seem to be necessary in defining the ordinal and metrical properties of points, straight lines, and planes (see below, p. 138). The "defect of starting point" alluded to in the passage is nothing more serious than an inappropriate logical arrangement of definitions, in which durations are used to define points (and also flat geometrical entities) instead of being defined by them.

After the above discussion it may come as something of a surprise to realize that the concept of a duration does in fact play an important role in the philosophy of organism (see below, p. 141). Clarity of understanding at this point requires the recognition of two facts: first, durations are no longer defined (as in the early works) in terms of direct perception but rather in causal terms; second, as defined in causal terms durations are not suitable for general use in extensive abstraction. I proceed to amplify these two statements. The concept of a duration as a complete nexus[17] of contemporary, or causally independent, occasions (PR, p. 192) is, on the face of it, quite different from the older concept of a duration as a single event. A possible interpretation which would bring the old and the new concepts of a duration quite close together should, however, be mentioned. This interpretation requires (*a*) that durations (as defined in PR) be a species of events (as defined in PR), and (*b*) that events (as defined in PR) be identical with events in the early works. I find (*a*) plausible, (*b*) rather less so, for the following reasons. An event is defined in PR as "a nexus of actual occasions inter-related in some determinate fashion in some extensive quantum. . . . One actual occasion is a limiting type of event" (p. 124). Hence durations (in the PR sense) are indeed events (in the PR sense) of a special kind. On the other hand, there is nothing very obvious about the identification of events in the old and new senses. Thus, for example, there are no minimum events (in the old sense), while a single actual occasion constitutes a minimum event (in the new sense). Also, one of Whitehead's examples of an event in PR is a molecule, whereas in his early books a molecule is regularly classified as a scientific object rather than as an event. None of these considerations is perhaps decisive against the identification of the two concepts of events, and a deeper study of PR than is possible here might actually support the identifica-

---

[17] ". . . a nexus is a set of actual entities in the unity of the relatedness constituted by their prehensions of each other, or—what is the same thing conversely expressed—constituted by their objectifications in each other" (PR, p. 35).

tion. Provisionally, however, I should prefer to distinguish the two concepts. (On events cf. also above, n. 15.)

Since durations (in the PR sense) are defined in causal terms, they are, in this respect, unlike bounded regions: the latter are always in principle prehensible (by means of coordinate or regional feelings) or at least, in the case of future regions, can be understood as extrapolations from prehensible regions; the former are only prehensible (and then indirectly) in quite exceptional cases. These exceptional cases occur when a duration is "presented" to an actual occasion by means of the close association of the duration with a region perceived in the mode of presentational immediacy. Such close association depends upon the existence of straight lines in the extensive continuum and this may well be a special feature of our present cosmic epoch (see below, p. 143). Thus, to admit durations as primitive entities in the generalized method of extensive abstraction formulated in PR would defeat what I take to be Whitehead's purpose of basing the method solely on extensional data prehensible, in principle, by *any* actual occasion in *any* cosmic epoch. On the other hand, the use of (presented) durations (as in PNK and CN) for deriving the four-dimensional space-time manifold of the present cosmic epoch would seem to be legitimate and even necessary (see below, p. 138).

2. Points and Straight Lines

Returning to the systematic development of the method of extensive abstraction, we must first define "non-tangential inclusion" which in turn will be used in defining abstractive sets. That one *can* define non-tangential inclusion solely on the basis of the assumptions and definitions already introduced is precisely the great advantage of starting with extensive connection (rather than extension) as a primitive concept.

Two regions *overlap* when there is a third region which they both include. It is assumed that overlapping is symmetrical; that if one region includes another, the two regions overlap; and that two regions which overlap are connected (Figure 18, iv). Two regions are *externally connected* when (1) they are connected, and (2) they do not overlap (Figure 18, v–vi). One region B is *tangentially included* in another region A when (1) B is included in A, and (2) there are regions which are externally connected with both A and B (Figure 18, ii–iii). One region B is *non-tangentially included* in another region A when (1) B is included in A, and (2) there is no third region which is externally connected with both A and B (Figure 18, i).

By defining an *abstractive set* in terms of the relation of non-tangential inclusion, we in effect limit abstractive sets to those called "simple" in our earlier discussion (cf. p. 64). Abstractive sets, as before, can *cover* one another and be *equivalent* to one another. A *geometrical element* is a complete group of abstractive sets equivalent to each other but not equivalent to any abstractive set outside the group. Each abstractive set belongs to one, and only one, geometrical element. The geometrical element *a* is *incident* in the geometrical element *b* when every member of *b* covers every member of *a*, but *a* and *b* are not identical. We now define a *point* as a geometrical element such that there is no geometrical element incident in it. (A "point" here has obvious affinities with the earlier "event-particle.") Points are in a sense absolute primes (the concept of a *prime* geometrical element in PR is essentially that of a prime abstractive class in CN —see Appendix I). A geometrical element prime with respect to the condition that the points *P* and *Q* are incident in it is called a *segment* between the points *P* and *Q*, which are known as the *end-points* of the segment. There are many segments with the same end-points, but a given segment has just a single set of end-points. The fact that there can be many segments with the same end-points is an example of the general principle that there can be many geometrical elements prime with respect to the same set of conditions. Especially important sets of conditions, however, will be those which define a unique geometrical element (e.g., the "straight" segment between two end-points); in fact, according to Whitehead, "The whole theory of geometry depends upon the discovery of conditions which correspond to one, and only one, prime geometrical element" (PR, p. 458).

A point is *situated* in a region when that region is a member of one of the abstractive sets which compose the point. A point is *situated in the surface* of a region when all the regions in which it is situated overlap that region but are not included in it. We now define a *complete locus* of points as a set of points composing either (1) all the points situated in a region (*volumes*) or (2) all the points situated in the surface of a region (*surfaces*) or (3) all the points incident in a geometrical element. A complete locus is assumed always to consist of an infinite number of points. When the geometrical element is a segment, the complete locus of points incident in it is a linear stretch. The order of points on a linear stretch can be specified by means of axioms for "between" (see, e.g., Axiom iv on p. 60).

Before sketching Whitehead's procedure for defining "straight" lines and, more generally, "flat" geometrical elements, I must emphasize the fact that Whitehead is not trying to define "straightness" or "flatness" in the *metrical* sense—an impossible task in view of what I take to be his starting point (namely, the purely topological concepts of regions and extensive connection). Thus, when Whitehead says that he has refuted "a dogma of science that straight lines are not definable in terms of mere notions of extension" (PR, p. 194), he must not be interpreted as referring to metrical straight lines (e.g., the straight lines of Euclidean geometry), but rather to the non-metrical straight lines of topology and projective geometry. This point will be discussed in some detail below.

Since my primary aim is to convey an intuitive understanding of the devices which Whitehead employs to arrive at his definitions of flat geometrical elements, I shall occasionally rely upon the use of diagrams to explain "visually" some of Whitehead's preparatory definitions. The precise definitions of all terms can, of course, be found in Part IV, chapters ii and iii of PR. We begin with the concept of "intersection" between regions. When two regions A and B overlap, their *intersection* (i.e., the region common to both A and B) may be *unique* or *multiple* (Figure 20, i–ii).

We come now to the crucial notion in the entire development, that of an "oval" region.[18] It is impossible to characterize a single oval in terms of White-

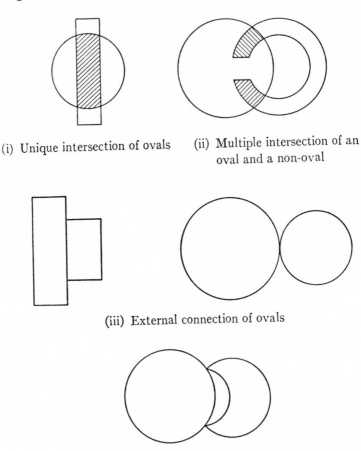

(i) Unique intersection of ovals   (ii) Multiple intersection of an oval and a non-oval

(iii) External connection of ovals

(iv) External connection of an oval and a non-oval

Fig. 20.—Intersection and external connection of oval and non-oval regions

[18] The term "oval" suggests "convex" and we shall soon see that convexity is indeed what Whitehead seems to have had in mind. The usual procedure in mathematics is to define convexity in terms of straightness, e.g., "A region is said to be convex if the [straight] interval joining any two points of it is composed entirely of points of the region" (O. Veblen, "The Foundations of Geometry," p. 20). In a certain sense, then, Whitehead is reversing the usual procedure by defining straightness in terms of convexity. However, it must be emphasized

head's two primitive concepts, but one can define a class consisting exclusively of ovals (though not necessarily of *all* ovals, Whitehead remarks). The definition of an *ovate class of regions* contains two sets of conditions (see PR, pp. 463–64) which I shall now attempt briefly to explain. There are nine "non-abstractive" conditions falling into three distinct groups, which may conveniently be called (1) intersection conditions (i–iii), (2) external connection conditions (iv–vi), and (3) existence conditions (vii–ix). The *existence conditions* require little discussion here: they correspond closely to some of our earlier axioms for extension, and their purpose is to insure the existence of the appropriate number and distribution of regions for constructing abstractive sets throughout the extensive continuum. Condition vii, however, is imperfectly formulated and deserves some consideration. In the "Corrigenda" at the end of PR Whitehead remarks that *"Condition vii* has been expressed carelessly, so as to apply only to the case of infinite spatiality, *i.e.*, to Euclidean and Hyperbolic Geometry" (p. 546). The trouble is, of course, that if space is elliptic (and therefore finite but unbounded), there will be finite sets of regions which "exhaust" the entire space and hence are not jointly included in any region (since the entire space, being unbounded, cannot itself be considered a region). Perhaps one might take account of the possibility of elliptic space and still fulfil the purpose of Condition vii—which is, roughly speaking, to guarantee that all regions belong to a single manifold—by substituting for Condition vii something like this: For any two regions B and C there is a finite sequence of regions belonging to the ovate class, $A_1, A_2, \ldots A_n$, such that B is connected with A_1, A_1 with $A_2, \ldots$, and A_n with C. Incidentally, the same difficulty with Condition vii arises if space is infinite but time is finite and unbounded; however, I find no evidence that Whitehead ever took this possibility seriously. A closed serial order for time would imply a cyclical evolution of the universe, with the possibility of, literally, a return to past states of the universe. The emphasis on the creativity of process in Whitehead's metaphysics would be hard to reconcile with such a theory of time.[19]

The *intersection conditions* are as follows: Any region in the ovate class can overlap another region in the ovate class only with unique intersection (i), but it must overlap some region outside the ovate class with multiple intersection (iii); and every region outside the ovate class must overlap some region in the ovate class with multiple intersection (ii).

The *external connection conditions* are as follows: Any region in the ovate class

---

that convexity as usually defined is a *metrical* concept and therefore requires a metrical concept of straightness in its definition. Whitehead's ovateness and straightness, as I have urged all along, are *non-metrical*.

[19] It is of considerable interest that there exist cosmological solutions of Einstein's gravitational field equations in which closed time-like lines occur. See K. Gödel, "An Example of a New Type of Cosmological Solutions of Einstein's Field Equations of Gravitation" and "A Remark about the Relationship between Relativity Theory and Idealistic Philosophy."

can be externally connected with another region in the ovate class only in such a way that their surfaces meet either in a complete locus of points or in a single point (iv); but it must be externally connected with some region outside the ovate class in such a way that their surfaces meet in a set of points which do not form a complete locus (vi); and any region not in the ovate class must be externally connected with some region in the ovate class in such a way that their surfaces meet in a set of points which is not a complete locus (v).

We come now to the group of three "abstractive conditions" (see PR, p. 464). These conditions, in effect, make the extensive continuum four-dimensional *relative to* the particular ovate class of regions being defined. We now assume that the extensive continuum of the present cosmic epoch possesses at least one ovate class satisfying the non-abstractive and abstractive conditions. Even though Whitehead holds it probable that there is only one such class in an extensive continuum, he does not succeed in proving this nor, he says, is the proof necessary for his argument—statements which are extremely perplexing and to which we shall return later on.

We define an *ovate abstractive set* as an abstractive set all of whose members belong to a given ovate class. A most important uniqueness theorem can now be proved, namely, that if two abstractive sets are prime with respect to the same twofold condition, (1) of covering a *given* group of points, and (2) of being equivalent to *some* ovate abstractive set, then they are equivalent. Whitehead's proof of this theorem goes as follows (PR, p. 465). If the two abstractive sets are equivalent to the same ovate abstractive set, the required conclusion follows immediately. Suppose the two abstractive sets are equivalent respectively to different ovate abstractive sets. Now, any pair of regions belonging respectively to these two ovate abstractive sets must intersect, since the given group of points is incident in all such regions. Also, the intersects in question must themselves be ovals in which the given group of points are incident. Hence, one can construct from these intersects a third ovate abstractive set which is covered by each of the two original ovate abstractive sets. But owing to the primeness of the two latter abstractive sets, they must both be equivalent to the third ovate abstractive set and hence equivalent to each other.

It follows that the group of abstractive sets satisfying a twofold condition of this type belong to a single geometrical element. For the case of two points, the uniqueness theorem yields the geometrical element which we call a *straight* segment; additional points yield *flat* geometrical elements of two and three dimensions. The locus of points incident in a straight segment is the *straight line* between the end-points of that segment. It is assumed that no two sets of a finite number of points, both in their *lowest terms* (i.e., such that no subset would define the same geometrical element as the set itself), define the same flat geometrical element. It follows immediately that two distinct straight lines can have at most one point in common. Also, *complete straight lines* (corresponding to the ordinary straight lines of geometrical theories) are now definable, as follows:

A complete straight line is a locus of points such that, (i) the straight line joining any two members of the locus lies wholly within the locus, (ii) every sub-set in the locus, which is in its lowest terms, consists of a pair of points, (iii) no points can be added to the locus without loss of one, or both, of the characteristics (i) and (ii) [PR, p. 466].

Planes are definable in analogous fashion (see PR, pp. 466–67).

It is time to evaluate the significance of the formal developments we have been following. I shall attempt to explain with the aid of some diagrams what I take to be the most plausible interpretation of Whitehead's accomplishments as he himself seems to have viewed them. Let us begin by examining some illustrations of the intersection and external connection of ovals and non-ovals (see Figure 20). Certain possibilities are, for simplicity, omitted in Figure 20; thus, e.g., two non-ovals (or an oval and a non-oval) may obviously intersect in a unique region; and two non-ovals (or an oval and a non-oval) may obviously be externally connected along a boundary which is a complete locus of points. But these possibilities do not constitute a violation of our non-abstractive conditions, which require only that *all* overlapping ovals have unique intersects; that every non-oval overlap *some* oval in multiple intersects; and that every oval overlap *some* non-oval in multiple intersects; and similarly for external connection.

One marked shortcoming of our diagrams for ovals and non-ovals is their two-dimensionality: in fact, the non-abstractive conditions depicted in Figure 20 are independent of any assumption as to the dimensionality of the regions (such an assumption is embodied in the abstractive conditions). However, the most misleading feature of our diagrams is their Euclidean character. In other words, what we find depicted in Figure 20 are *Euclidean* ovals and non-ovals, but it is evident that nothing that has been assumed so far could possibly single out such a *metrically* characterized class of regions. Thus, if Figure 20 is to be an accurate representation of the intersection and external connection conditions for ovals, one must imagine each region depicted to stand for *the entire class of regions topologically equivalent to it.* But this implies that there are indefinitely many ovate classes of regions in the extensive continuum; for, given some definite ovate class of regions throughout the extensive continuum (e.g., the Euclidean one of our diagrams), it is always possible to find another ovate class of regions throughout the extensive continuum by imagining that continuum more or less deformed, while preserving all intersection and external connection properties of the original class of regions. And there is nothing in our definition of an ovate class which forbids such a deformation. We see now the strangeness of Whitehead's doubts concerning the number of ovate classes in the extensive continuum: *as ovate classes are defined,* the extensive continuum *must* contain an infinity of them, although, of course, additional conditions may be imposed which serve to reduce the number of ovate classes, perhaps even to one.

Corresponding to the infinity of different ovate classes one would expect to

find an infinity of different meanings for "straight lines" as Whitehead defines
them. To see that this expectation is indeed correct, let us examine a diagram
which illustrates Whitehead's definition of a straight segment (see Figure 21).
We note that there are possible non-ovate abstractive sets which converge to
straight segments; but while all ovate abstractive sets (satisfying the appropriate conditions) converge to straight segments, not all non-ovate abstractive

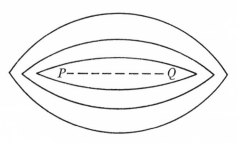

(i) An ovate abstractive set "converging" to the "straight" segment *PQ*

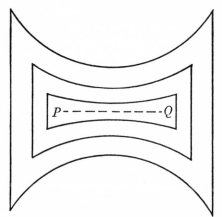

(ii) A non-ovate abstractive set "converging" to the "straight" segment *PQ*

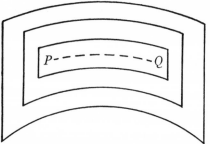

(iii) A non-ovate abstractive set "converging" to the "curved" segment *PQ*

FIG. 21.—The relation between ovateness and straightness

sets (satisfying the appropriate conditions) converge to straight segments. Now, topologically, i, ii, and iii in Figure 21 are indistinguishable, i.e., any one of them can be obtained from any one of the others by a topological transformation. Generalizing, we may say that just as there was an infinity of ovate classes of regions in the extensive continuum, so, *as straight lines are defined*, there must be an infinity of families of such lines in the extensive continuum (in fact, there is a one-one correlation between the ovate classes and the families of straight lines). Three of these families of straight lines will correspond to the familiar geodesics (or shortest distances between any two points) of the three types of metrical geometry. But there will be many other families of straight lines that correspond to nothing that would ordinarily be called "straight"—at least not "straight" for sense-perception (e.g., S-shaped curves, spirals, etc.). Once again, additional conditions may be imposed to single out just one or a few families of straight lines from among all the rest; this in fact seems to be Whitehead's procedure, as we shall see in the next section.[20]

Whitehead considers his definition of a straight line to be an improvement upon Euclid's ("A straight line is any line which lies evenly with the points on itself") on several grounds. As we have seen, Whitehead's definition does not take the notion of "evenness" for granted and also it guarantees the uniqueness of the straight line between any two points. It will be recalled that the uniqueness property depends only on the intersection properties of ovals; while the complete loci which form the common boundaries of externally connected ovals seem to exemplify the property which the Greek geometers referred to by the term "even": "On either side of such a locus, there is the interior of one oval and the exterior of another oval, so that the locus is 'even' in respect to the contrasted notions of 'concavity' and 'convexity' " (PR, p. 468).[21] However, that flat loci and even loci are identical is "an extra 'assumption'—provable or otherwise according to the particular logical development of the subject which may have been adopted . . ." (PR, p. 468). An alternative to Euclid's definition is to define a straight line as "the shortest distance between two points." Whitehead prefers his own definition because it is independent of distance and therefore of measurement, which process itself, according to Whitehead, presupposes straight lines.

The allusion to Euclid's definition of a straight line should not mislead one as to the character of Whitehead's definition of a straight line: it is *not* possible to deduce metrical properties of straight lines (e.g., that a straight line is the shortest distance between two points) from Whitehead's definition, since distance (the key idea of metrical geometry) has not been introduced. The whole point of Whitehead's definition, as we shall see more fully in the next section, is

---

[20] It is instructive to compare Whitehead's procedure with that of E. V. Huntington in his articles "A Set of Postulates for Abstract Geometry, Expressed in Terms of the Simple Relation of Inclusion" and "The Duplicity of Logic." See Appendix II for such a comparison.

[21] The concept of externally connected regions also serves to explain the notion of "continuous transmission" in physics without running afoul of Zeno's paradoxes (see PR, p. 468).

that it applies equally well to Euclidean and non-Euclidean geometries. In fact, Whitehead lists as the most important properties of straight lines which follow from his definition only the following: "(i) their completeness, (ii) their inclusion of points, (iii) their unique definition by any pair of included points, (iv) their possibility of mutual intersection in a single point" (PR, p. 503). I conclude that Whitehead's straight lines are essentially nothing more than a certain class of what are usually called "topological lines," namely, those satisfying certain conditions of intersection and inclusion of points. The significance of just these conditions will now be discussed.

3. Points and Straight Lines in Projective and Metrical Geometry

To appreciate the importance which Whitehead attaches to his definition of a straight line in PR and to evaluate the claims made for that definition, one must be familiar with certain nineteenth-century developments[22] in geometry which form the essential background to Whitehead's own investigations. Repeatedly, in the discussions of geometry in his various writings, Whitehead refers with high praise to the work of K. G. C. von Staudt and A. Cayley. (See, e.g., PR, p. 505.) It will be well, therefore, to begin by briefly recounting the relevant contributions of these—and a few other—nineteenth-century mathematicians.

A new branch of geometry called "projective geometry" was initiated around the beginning of the nineteenth century, mainly by the French mathematicians Poncelet and Chasles and the German mathematicians Steiner, von Staudt, Möbius, and Plücker. What especially distinguishes this new geometry from traditional Euclidean geometry is the complete absence of any reference to concepts such as angle, distance, parallelism, or congruence. Instead, projective geometry deals with those properties of a figure which remain unchanged when the figure is "projected" into (in general) some other figure. The nature of the process called projection may easily be visualized as follows. Given any figure F_1 in a plane and any point V outside this plane. Now let V, the "center of projection," be joined by a straight line to every point $A_1, B_1, C_1 \ldots$ of F_1, and let these lines intersect a second plane in the points $A_2, B_2, C_2, \ldots$ This last set of points defines a new figure F_2, which is said to be the projection of the original figure F_1. Notice that although F_1 and F_2 may look quite different when placed side by side, to an observer whose eye is at V (and lacking all depth-vision) F_1 and F_2 would be indistinguishable. Projectively, then, all triangles regardless of size and shape are equivalent, all conic sections are equivalent, and so forth. As a very simple illustration, consider any projection of a circle in a given plane into a second plane neither parallel nor perpendicular to the

---

[22] An elementary introduction to these developments is available in H. G. Forder, *Geometry*. Further details may be found in F. Klein, *Elementary Mathematics from an Advanced Standpoint*, Vol. II, *Geometry*, and G. deB. Robinson, *The Foundations of Geometry*.

first plane; the projected figure will be an ellipse. In general, it follows immediately from the definition of a projection that it is a transformation of figures into other figures such that points become points, straight lines become straight lines, straight lines intersecting in a point become straight lines intersecting in a point, and points on a single straight line become points on a single straight line.[23] Thus, the characteristic mark of the theorems of projective geometry is that they are concerned solely with the mutual incidence of points, straight lines, and planes.

The discussion of projective geometry up to this point has relied heavily on an assumed familiarity with Euclidean geometry—our illustrations, for example, have presupposed for terms such as "straight line," "plane," "triangle," and "conic section," their familiar Euclidean meanings. It cannot now be too strongly emphasized that no such Euclidean interpretation is to be read into the fundamental concepts and principles of projective geometry (in fact, as we shall see, the principles of projective geometry hold also in non-Euclidean geometries). The best way of avoiding illegitimate appeals to intuition in the interpretation of an abstractly formulated subject matter is to adopt the axiomatic approach, where the primitive concepts are uninterpreted and possess just so much meaning as is given them by the axioms in which they occur. Let us then briefly examine the nature of projective geometry considered as an axiom-system. There are, of course, many alternative, essentially equivalent, axiomatic formulations of projective geometry; no one of these formulations will be described here in detail but the discussion will be guided by Whitehead's formulations (in APG and AG).

The fundamental primitive concepts are the entities termed *points, straight lines,* and *planes* (the last may also be defined in terms of the first two); and the relations termed *containing* (or its converse *lying in*) and *intersection.* Among the fundamental axioms of projective geometry[24] are two distinct sets. (1)

---

[23] Analytically, a projective transformation of coordinates (in three dimensions) may be characterized as a transformation in which x', y', z' are fractional linear functions, with the same denominator, of x, y, z:

$$x' = \frac{a_1x + b_1y + c_1z + d_1}{a_4x + b_4y + c_4z + d_4},$$

$$y' = \frac{a_2x + b_2y + c_2z + d_2}{a_4x + b_4y + c_4z + d_4},$$

$$z' = \frac{a_3x + b_3y + c_3z + d_3}{a_4x + b_4y + c_4z + d_4}.$$

It must be emphasized, however, that much of projective geometry can be developed without any use of coordinates at all. When coordinates are used in pure projective geometry, their significance is merely *ordinal,* i.e., they are names for referring to the relative order of points on a line.

[24] See, for example, Whitehead's own tract on the subject, APG, as well as his *Encyclopaedia Britannica* article, AG. A less formal presentation may be found in B. Russell, *An Essay on the Foundations of Geometry,* chap. iii. Reference may also be made to the books cited in n. 22.

Axioms of Classification (or Connection): e.g., points form a class of entities with at least two members; any two distinct points lie in one and only one straight line; there is at least one straight line which does not contain all the points. (2) Axioms of Order: e.g., if A, B, C are three distinct collinear points, and D lies in the segment ABC, then the segment ADC is contained within the segment ABC (this axiom presupposes a definition of the *segment* determined by three points); there exists at least one straight line for which the point order possesses the Dedekind property.

From any of the customary sets of Axioms of Classification one can deduce the following. (1) Two distinct points determine a unique straight line. (2) Two distinct coplanar straight lines always intersect in a unique point. (3) Three distinct points determine a unique plane. (4) Two distinct planes always intersect in a unique straight line. (5) The straight line containing any two points in a plane lies wholly in that plane. (6) The projective space has three dimensions.

From any of the customary sets of Axioms of Order one can deduce the fundamental properties of a *closed serial* (or *cyclic*) *order* for the points on a straight line. Such an order is characteristic of, for example, the points on a circle; that it is the appropriate order for projective geometry can be seen from the following considerations. The relation of three collinear points such that one *lies between* the other two is not in general preserved by a projection. Thus, if B lies between A and C before projection, it may lie between C and A after projection, a fixed "sense" (say, from left to right) for the order of points on a line having been chosen. On the other hand, there is a relation of four collinear points such that it is always preserved by a projection. This relation is called *separation* (of one pair of points by another pair), and it may be intuitively characterized by saying that the pairs of points AC and BD separate each other if and only if the triads (ACB) and (ACD) have different senses (say, one clockwise and the other counterclockwise). Since the relation of B lying between A and C has no definite meaning for three points on a circle while the relation of separation for four points on a circle has a definite meaning, we see that a closed serial order is indeed the proper order for points on a line in projective geometry.

We come now to the "fundamental theorem of projective geometry." First, we note that a *projective transformation* can easily be defined in terms of our primitive concepts. The theorem in question then states that a projective transformation between the points on two straight lines is completely determined when the correlates of three distinct points on one line are determined on the other. In constructing an axiom-system for projective geometry, one usually adopts the fundamental theorem or some essentially equivalent one (such as Pappus' theorem)[25] as an Axiom of Classification. However, there is an alternative procedure (due to von Staudt) which Whitehead follows, namely, one de-

---

[25] Pappus' theorem states: If l and l' are two distinct coplanar lines, and A, B, C are three distinct points on l, and A', B', C' are three distinct points on l', then the three points of intersection of AA' and $B'C$, of $A'B$ and CC', of BB' and $C'A$, are collinear.

duces the fundamental theorem from the Axioms of Classification (without Pappus' theorem or any equivalent theorem) *together with* the Axioms of Order (see AG, p. 183, and APG, p. 33).

The nineteenth century saw the development of non-Euclidean metrical geometries, at first quite independently of projective geometry. Lobatchewsky and Bolyai independently formulated a geometry (now called *hyperbolic* geometry) in which through any given point two straight lines may be drawn parallel to a given straight line. In this case there will be an infinity of straight lines through the given point which do intersect, and an infinity of straight lines through the given point which do not intersect, the given straight line. Riemann formulated a geometry in which there are no straight lines through a given point and parallel to a given straight line. This type of geometry departs from Euclidean geometry even more radically than does hyperbolic geometry. In the latter straight lines are, as in Euclidean geometry, infinite in length, whereas in Riemann's geometry straight lines are finite in length. There are two distinct forms of Riemann's geometry: in his original form (now called *double elliptic* or *spherical* geometry) two straight lines always intersect in a pair of points; in the form later discovered by Klein (now called *single elliptic* or simply *elliptic* geometry) two straight lines always intersect in a single point. Elliptic, but not spherical, geometry obviously resembles both Euclidean and hyperbolic geometry in that two straight lines can intersect in, at most, a single point. A profound connection has been shown to exist between elliptic and projective geometry, namely, projective geometry is simply elliptic geometry with all metrical relations omitted. Spherical geometry in two dimensions is well known; it is the geometry on the surface of a sphere. Also, by identifying every pair of antipodal points on the surface of a sphere, one obtains a model for two-dimensional elliptic geometry.

Riemann also undertook more general investigations into the foundations of geometry, his fundamental conceptions being that of a *manifold* (discrete or continuous) and that of the *measure of curvature* (of a continuous manifold).[26]

---

[26] B. Riemann, "Über die Hypothesen, welche der Geometrie zu Grunde liegen"; translated into English by W. K. Clifford, *Mathematical Papers*, pp. 55–71. Riemann's measure of curvature is in a sense a generalization of Gauss's definition of curvature for two-dimensional surfaces. Consider any two-dimensional surface on which the line-element ds is given by the following expression:

$$ds^2 = E du^2 + 2F du\, dv + G dv^2 ,$$

where u and v are arbitrary curvilinear coordinates, of which E, F, and G are functions. Then the Gaussian curvature K may be calculated as a function of u and v from a differential expression of the second order in E, F, and G. In its explicit form the expression for K is rather complicated (see F. S. Woods, "Forms of Non-Euclidean Space," p. 38, or D. J. Struik, *Lectures on Classical Differential Geometry*, p. 112). A considerable simplification is introduced if one adopts orthogonal coordinates, since in this case F vanishes; one then has:

$$K = - \frac{1}{\sqrt{(EG)}} \left[\frac{\partial}{\partial u} \left(\frac{1}{\sqrt{E}} \frac{\partial \sqrt{G}}{\partial u} \right) + \frac{\partial}{\partial v} \left(\frac{1}{\sqrt{G}} \frac{\partial \sqrt{E}}{\partial v} \right) \right] .$$

By equating the expression for K to a constant one obtains a second order differential equation whose solution is given by equation (2). The expression corresponding to K for three- (or

To make measurement of length possible in a continuous manifold, Riemann assumes that lengths of lines throughout the manifold must be comparable with one another. Given an n-dimensional manifold so that the (arbitrary curvilinear) coordinates of a point may be represented by n real numbers $x_1 \ldots x_n$, various general considerations about the nature of length lead Riemann to postulate that ds, the infinitesimal increment of length, should be a linear homogeneous function of $dx_1 \ldots dx_n$, of the following form:

$$d s^2 = \sum_{j=1}^{n} \sum_{i=1}^{n} g_{ij} d x_i d x_j, \qquad (1)$$

where the g_{ij} are in general functions of $x_1 \ldots x_n$. A further specialization of ds for the case of what may be called a "homogeneous isotropic" manifold results when Riemann assumes that figures are to be freely movable throughout the manifold, which is equivalent to the assumption that the measure of curvature is constant for all points and all directions at each point;[27] ds^2 can then be written:

$$d s^2 = \frac{\sum\limits_{i=1}^{n} d x_i^2}{\left[1 + \dfrac{K}{4} \sum\limits_{i=1}^{n} x_i^2 \right]^2}, \qquad (2)$$

where K is the measure of curvature. If K is negative, one obtains infinite hyperbolic space; if K is positive, one obtains finite but unbounded elliptic space; if K is zero, one obtains ordinary infinite Euclidean (or *parabolic*) space.

The work of Riemann to which I have just referred is, according to Whitehead, open to criticism or development in various ways, of which the one[28] that particularly concerns us here is that "the introduction of co-ordinates is entirely unexplained and the requisite presuppositions are unanalysed" (NG, p. 220). By this Whitehead apparently means that Riemann gives no proof that the order (suitably defined) of points on a line can be represented by the order

---

higher-) dimensional surfaces is considerably more complicated; it is called the Riemann-Christoffel curvature tensor and plays an important role in Einstein's general theory of relativity (see chap. viii and Appendix III).

[27] A theorem due to F. Schur shows that if the curvature is constant for all directions at each point, then it is also constant from point to point; or, in other words, if a manifold is isotropic then it is also homogeneous.

[28] The other three are: "the idea of a manifold requires more precise determination" (NG, p. 220), in particular, the meaning of continuity must be defined—as was done by Dedekind and Cantor; "the assumption that ds is the square root of a quadratic function of $dx_1, dx_2 \ldots$ is arbitrary" (NG, p. 220)—a proof was given by Helmholtz and Lie; "the idea of superposition, or congruence, is not adequately analysed" (NG, p. 220)—this was accomplished by Lie (see below, p. 134).

For Whitehead the study of manifolds satisfying equation (1) is not "geometry" in his sense at all. On this point see below, p. 189.

of the real numbers. The great contribution of von Staudt was to show how to introduce numerical coordinates into projective geometry. It is worth quoting Whitehead's judgment on this achievement: ". . . numerical coordinates, with the usual properties, can be defined without the introduction of distance as a fundamental idea. The establishment of this result is one of the triumphs of modern mathematical thought. It has been achieved by the development of one of the many geometrical conceptions which we owe to the genius of von Staudt" (APG, p. v). It is unnecessary for our purposes to discuss the details of von Staudt's conception, but I shall attempt to sketch the general procedure.

We begin with the fundamental definition of projective geometry (due to von Staudt), that of *harmonic conjugates:*

Harm. ($ABCD$) symbolizes the following conjoint statements: (1) that the points A, B, C, D are collinear, and (2) that a quadrilateral can be found with one pair of opposite sides intersecting at A, with the other pair intersecting at C, and with its diagonals passing through B and D respectively. Then B and D are said to be "harmonic conjugates" with respect to A and C [AG, p. 180].

It can be shown on the basis of the classificatory axioms of projective geometry (exclusive of the fundamental theorem) that if A, B, C be any three distinct collinear points, there exists a unique point D such that Harm. (A, B, C, D). A *harmonic sequence* of points on any straight line may now be introduced. Beginning with three distinct points P_0, P_1, P_∞ on a line l, one can find a unique point P_2 on l such that Harm. ($P_0P_1P_2P_\infty$). Next, another point P_3 can be found such that Harm. ($P_1P_2P_3P_\infty$), and this process may be continued indefinitely *provided* we know that P_3, . . . P_4, . . . are all distinct points not identical with P_0, P_1, P_2, or P_∞. Specifically, the existence of the requisite infinity of points can be assured by an appropriately chosen classificatory or ordinal axiom.[29] A set of points obtained in this way is the harmonic sequence on the line l.

In order to introduce numerical coordinates, von Staudt made use of a projective geometrical construction for defining the *sum* and *difference* of any two points (relative to two fixed points, O and U) and another projective geometrical construction for defining the *product* and *quotient* of any two points (relative to three fixed points, O, U, and I).[30] It is important to note that the construction

[29] The classificatory axiom, due to Fano, may be stated in the form: "Harm. ($ABCD$) implies that B and D are distinct points," or equivalently, "Not every harmonic sequence on a line contains a finite number of points." The ordinal axiom may be stated in the form: "For any three collinear points A, B, C, there is a point D such that the pairs of points AC and BD separate each other." For a discussion of this matter see Robinson, *The Foundations of Geometry*, p. 123.

[30] To talk of "adding" and "multiplying" points means, of course, that the *formal* properties of the familiar algebraic operations of addition and multiplication also characterize the geometrical constructions (e.g., the constructions for adding and multiplying two points are commutative, associative, and distributive). Similarly, the points O, I, U have the formal properties of the numbers zero, one, and infinity respectively. For details see Robinson, *The Foundations of Geometry*, pp. 91–94 or Forder, *Geometry*, pp. 74–75.

for addition and subtraction presupposes all of the classificatory axioms except the fundamental theorem (or Pappus' theorem), while the construction for multiplication and division presupposes all of the classificatory axioms including the fundamental theorem (or Pappus' theorem). In these constructions it is found that O plays the part of "zero," I plays the part of "unity," and U plays the part of "infinity." We now identify the points O, I, U respectively with the three points, P_0, P_1, P_∞ employed earlier in constructing a harmonic sequence on a line. And by applying our previously defined constructions for addition, subtraction, multiplication, and division of points, we obtain from P_0, P_1, P_2, P_3, . . . P_∞ a new set of points which may be designated $P_{m/n}$ (where m and n are any two positive or negative integers). This totality of points is called the *harmonic net* or *net of rationality* determined by P_0, P_1, . . . P_∞. Using the ordinal axioms, we may correlate the points on any line with all the real numbers, positive and negative. The entire procedure can be extended to the plane and then to three dimensions.

It can now be shown that, on the basis of a coordinate-system of the kind introduced above, straight lines and planes are represented by linear equations and conic sections are represented by quadratic equations. Also, the important concept of "cross- (or anharmonic) ratio" can be defined. Given any range of four collinear points $(Q_1Q_2Q_3Q_4)$ with coordinates q_1, q_2, q_3, q_4, the cross-ratio of that range of points, $\{Q_1Q_2Q_3Q_4\}$, is defined as follows:

$$\{Q_1Q_2Q_3Q_4\} = \frac{(q_1 - q_2)(q_3 - q_4)}{(q_2 - q_3)(q_4 - q_1)}.$$

The equality of the cross-ratios of any two ranges of points is a necessary and sufficient condition for their mutual projectivity. Also, the cross-ratio of a harmonic range of points is always equal to -1.

Once numerical coordinates have been introduced into projective geometry, it is possible, by a method due originally to Cayley and elaborated by Klein, to define a numerical measure of the distance between any two points. At this point some criteria as to the adequacy of any such proposed definition would clearly be desirable. In UA, Whitehead formulates the following three axioms for the quantitative relation (called "distance") between any two points in a spatial manifold:

Axiom I. Any two points in a spatial manifold define a single determinate quantity called their distance, which, when real, may be conceived as measuring the separation or distinction between the points. When the distance vanishes, the points are identical.

Axiom II. If p, q, r be three points on a straight line, and q lie between p and r, then the sum of the distances between p and q and between q and r is equal to the distance between p and r.

Axiom III. If a, b, c be any three points in a spatial manifold, and the distances ab and bc be finite, then the distance ac is finite. Also if the distance ab be finite and the distance bc be infinite, then the distance ac is infinite. Also if the distances ab and bc be real, then the distance ac is also real [UA, pp. 349–50].

The Cayley-Klein procedure for defining distance[31] in projective terms depends upon an important property of the cross-ratio of a range of points. If A_1, A_2, P_1, P_2, P_3 are collinear points, it can be shown that

$$\{A_1P_1A_2P_2\}\{A_1P_2A_2P_3\} = \{A_1P_1A_2P_3\},$$

or,

$$\log\{A_1P_1A_2P_2\} + \log\{A_1P_2A_2P_3\} = \log\{A_1P_1A_2P_3\}. \tag{3}$$

If we now define the distance P_1P_2 between the two points P_1 and P_2 as some multiple of log $\{A_1P_1A_2P_2\}$, equation (3) guarantees the additive property of distance relative to collinear points. Thus our formal definition of distance becomes:

$$\text{dist}\,(P_1P_2) = \frac{\gamma}{2}\log\{A_1P_1A_2P_2\}, \tag{4}$$

where γ is an arbitrary constant.

Once γ is chosen we have a definite measure of distance along the line determined by A_1 and A_2. Clearly, the same definition of distance can be applied to every straight line in a plane, but the pair of points A_1, A_2 will differ from line to line, and we have as yet no way of comparing distances along different lines. What we need is some absolute configuration that will furnish once and for all exactly two points on every line. These points must obviously be given by some line, straight or curved, which intersects every straight line in the plane in two points. It can be shown that the only line satisfying this condition is a conic section. It must be remembered that the pair of points in which a given straight line intersects the conic section may be imaginary. Also, the conic section may degenerate into a pair of intersecting straight lines or into a pair of points with the straight line joining them taken twice. (Incidentally, in order to define a measure of angle throughout the plane we need a definite pair of straight lines passing through each point of the plane, and these lines are also furnished by the conic section in the form of the two tangents from any point to the conic section.)

Generalizing to the case of three dimensions, we thus introduce a quadratic surface, called *the absolute*, with the equation:

$$k\,(x^2 + y^2 + z^2) + 1 = 0 \qquad (k\text{ real}), \tag{5}$$

and define distance with respect to the absolute. In other words, we identify A_1 and A_2 in equation (4) with the pair of points in which the line determined by P_1 and P_2 cuts the absolute. The absolute is an imaginary quadratic surface for k positive, a real quadratic surface for k negative, and the plane at infinity together with an imaginary conic section in that plane for $k = 0$. It can be shown that the constant γ in equation (4) is so related to the constant k in equation (5) that when k is positive γ must be imaginary and when k is negative γ

[31] The other essential metrical concept, angle, can be defined by a procedure analogous to that for distance. An excellent simplified account of both procedures is given by C. A. Scott, "On Cayley's Theory of the Absolute."

must be real. Furthermore, γ is closely related to K, Riemann's measure of curvature in equation (2), viz., $K = -1/\gamma^2$; thus, imaginary and real values for γ give rise respectively to elliptic and hyperbolic geometry. Parabolic geometry emerges as a limiting case of either of the other two types when k approaches zero; this is ordinary Euclidean geometry with the expression for distance:

$$\text{dist} \ (P_1 P_2)^2 = (x_1 - x_2)^2 + (y_1 - y_2)^2 + (z_1 - z_2)^2 \ ,$$

where (x_1, y_1, z_1) and (x_2, y_2, z_2) are respectively the coordinates of the points P_1 and P_2.

Returning to Whitehead's criteria for a definition of distance, we see that Axioms II and III are obviously satisfied by the Cayley-Klein definition. Axiom I seems to be violated because the Cayley-Klein definition refers to *four* points instead of two. However, Whitehead argues (UA, p. 354) that since the two points on the absolute are always either both imaginary or both at infinite distance from all other real points, one has in effect defined distance as a function just of the arbitrary pair of real points P_1 and P_2. There remains for consideration the arbitrariness involved in the use of the absolute for defining distance—an arbitrariness which it is the great merit of Lie's theory of congruence-groups to have removed.[32]

In Whitehead's earliest published discussion of the Cayley-Klein definition of distance (in UA, written between 1890 and 1898), he completely ignores Lie's theory—thus the three axioms for distance say nothing about congruence (although the relevant papers by Lie are cited in the "Historical Note" to chap. i, Book VI, of UA). Subsequently, Whitehead treats Lie's theory in some detail (see ADG, chap. v, and AG, pp. 189–90). Here we shall have to be content with a few brief and loosely worded remarks. Lie's problem is to formulate a precise meaning for "free mobility" (the necessary condition for testing the congruence of figures by means of superposition). Lie's tool in solving this problem is the concept of a "group of motions," which may be explained as follows. The result of two or more successive motions (of a rigid body) is always equivalent to a single motion; but this is the characteristic property of groups in mathematics (which suggested to Lie the application of his previously elaborated theory of groups). Now, a motion (of a rigid body) may be thought of as a species of one-one point transformations, i.e., as a shift of each point of the body to one single other point. By extending this conception to one-one transformations of *all* points in space, one arrives at the notion of a complete group of motions, also called a congruence-group. In other words, "The displacement of a rigid body is simply a mode of defining to the senses a one-one transformation of all space into itself" (AG, p. 189).

It is obvious that a congruence-group must be a sub-group of the general

---

[32] Lie's work on the foundations of geometry was undertaken at the request of Klein, who wished Helmholtz' results on congruence and geometry to be placed on a firmer foundation; see Russell, *An Essay on the Foundations of Geometry*, p. 48.

projective group of transformations (since the motion of a rigid body will always transform a point into a point, a line into a line, and a point with a line through it into a point with a line through it). The really critical property in Lie's characterization of a congruence-group is what he calls "free mobility in the infinitesimal." It amounts to this: (1) any point can be moved into any other position; (2) if a point and an arbitrary line through it are fixed (i.e., each transforms into itself, or is "latent"), continuous motion is still possible; (3) if in addition to the point and the line through it an arbitrary plane through both of them is fixed (or latent), then no continuous motion is any longer possible.[33]

Lie goes on to prove that conditions (1), (2), and (3) are satisfied only for congruence-groups such that a quadratic surface of the form expressed by equation (5) (including the degenerate case when $k = 0$) is latent. Equation (5), or the absolute, thus enables us to connect Lie's theory of congruence-groups with the Cayley-Klein definition of distance. We note the existence of an indefinite number of congruence-groups (and therefore a correspondingly indefinite number of expressions for distance): one for each closed real quadratic surface (hyperbolic congruence-groups), one for each imaginary quadratic surface with a real equation (elliptic congruence-groups), and one for each imaginary conic in a real plane and with a real equation (parabolic congruence-groups). Furthermore, Lie's results prove that the alternative Cayley-Klein expressions for distance do indeed exhaust the possible definitions of distance which satisfy Whitehead's three axioms (see above, p. 132) and *also* the following additional axiom:

Axiom IV. The distance between any two points is invariable during a transformation of the congruence-group.

That Axiom IV holds for distance defined by equation (4) can be easily demonstrated. During any transformation of a given congruence-group the points A_1, A_2, P_1, P_2 will be transformed into A_1', A_2', P_1', P_2' in such a way that (a) $\{A_1P_1A_2P_2\} = \{A_1'P_1'A_2'P_2'\}$ because the transformation is projective, and (b) A_1', A_2' are on the absolute because A_1, A_2 are on it, and the absolute is latent.

It has already been mentioned that angular measure can be defined with respect to the absolute; also, parallelism has a particularly simple and illuminating interpretation: a pair of coplanar straight lines which intersect on the absolute are said to be parallel. The justification for this definition lies in the fact that the two tangents to the absolute from the intersection of such a pair of lines coincide, and it then turns out that the Cayley-Klein measure of angle gives zero for the angle between the pair of lines. But in ordinary metrical geometry, to say that the angle between the two coplanar lines is zero is equiva-

---

[33] Characteristically, Whitehead criticizes Lie's procedure because "The conception of a *finite* continuous group, though it is simple enough analytically, does not seem to correspond to any of the obvious and immediate properties of congruence-transformations as presented by sense-perceptions" (ADG, p. 47). Whitehead proposes instead of Lie's conception of a finite continuous group a set of axioms which "conform more closely to the obvious properties of congruence-transformations" (ADG, p. 47).

lent to saying that the two lines are parallel; hence the naturalness of the definition of parallelism. In terms of the preceding definition of parallelism, the distinction among the three kinds of metrical geometry can be expressed quite simply. If the absolute is an imaginary quadratic surface, then any pair of coplanar lines can intersect on the absolute only in an imaginary point. Thus, no real parallel lines exist, so that if the imaginary portion of projective space is excluded, the remaining real portion is identical with elliptic space (a result already mentioned above). On the other hand, if the absolute is a real quadratic surface, then for any given line l and any given point P not on l, there will be two lines coplanar with P and l which pass through P and are parallel to l, namely, the two lines joining P to the two points of intersection of l with the absolute. It can then be shown that the portion of projective space inclosed by the absolute in this case is identical with hyperbolic space, points on the boundary of the absolute corresponding to points at infinity. Finally, when the absolute degenerates into the plane at infinity together with an imaginary conic section in that plane, there will be but a single real point of intersection of any line with the absolute; hence, there will be exactly one line parallel to a given line l and passing through a given point P outside of l, since there is just one line through P and intersecting l on the absolute. Thus, if the plane at infinity is excluded, the remaining portion of real projective space is identical with parabolic (or Euclidean) space. This exclusion of the plane at infinity is, of course, just the reverse of the procedure often used for the construction of projective space from Euclidean space (exclusive of its metrical properties), namely, one supplements the ordinary Euclidean points and lines with "extraordinary" points and lines in the plane at infinity.

The preceding discussion may be summarized by means of the distinction between projective geometry and what Whitehead calls "descriptive geometry." I cite first Whitehead's general definition of "geometry" as "the theory of the classification of a set of entities (the points) into classes (the straight lines), such that (1) there is one and only one class which contains any given pair of entities, and (2) every such class contains more than two members" (AG, p. 179). *Descriptive geometry* is defined as any geometry in which two coplanar straight lines do not necessarily intersect and in which a straight line is an open series of points. (In projective geometry, as we have seen, two coplanar straight lines necessarily intersect and the straight line is a closed series of points.) Projective geometry is unique (more precisely, its axiom system is categorical), while there are many different descriptive geometries. A descriptive space (e.g., hyperbolic or Euclidean metrical space) can always be considered as a region within a larger projective space and, conversely, projective space can be considered as a suitably supplemented descriptive space. The notion of such supplementary entities (called *ideal elements*) is due to von Staudt, who introduced them into Euclidean geometry; the considerably more difficult problem of introducing ideal elements into hyperbolic geometry was solved by Klein and M. Pasch (see ADG, chaps. ii and iii, and AG, pp. 187–88).[34]

Perhaps the most striking thing about the results summarized above is the view they afford of the relation between projective geometry and metrical geometry. Before Cayley, projective geometry had seemed to be nothing but an impoverished metrical geometry, whereas Cayley showed how metrical geometry might be considered as a special case of projective geometry, whence his remark: "Projective geometry is all geometry."

4. The Philosophical Import of Straightness and Flatness

Where, now, do Whitehead's definitions of points, straight lines, and planes fit into the line of thought we have just been considering? Normally, as we have seen, one takes points and straight lines (and perhaps planes) as undefined elements in projective geometry. Whitehead, however, has succeeded in *defining* these three concepts in terms of what he calls "mere notions of extension" (which seem to be identical with what are usually called topological notions). Thus, Whitehead begins the formal development of his theory of extension in PR with two undefined notions—*region* and *extensive connection*—and on the basis of various assumptions involving ultimately only these two notions, he defines *point, straight line, plane,* and *three-dimensional flat space*. The definitions of the first three are such as to satisfy all the usual classificatory axioms of projective geometry—with one exception,[35] namely, the characteristic principle of projective geometry, that "two coplanar straight lines always have one point in common, and two planes always have one straight line in common." However, one can replace this principle with the weaker principle, characteristic of descriptive geometry, that "two coplanar straight lines always have one point, or none, in common, and two planes always have one straight line or else not a single point in common" (which does follow from Whitehead's definitions), and then supplement the defined points, lines, and planes with "ideal" or "projective" points, lines, and planes.

Before proceeding any further it must be pointed out that Whitehead's formal development of the theory of extension in PR ends with his definitions of straight lines, planes, and three-dimensional flat spaces. He does go on to discuss informally the essential non-metrical distinctions among the Euclidean, elliptic, and hyperbolic geometries and refers finally to the definition of congruence and distance in terms of non-metrical concepts (see PR, pp. 504–5).

---

[34] An *ideal point*, for example, may be defined as follows: ". . . an 'Ideal Point,' is the class of lines which is composed of two coplanar lines, *a* and *b*, say, and of the lines formed by the intersections of pairs of distinct planes through *a* and *b* respectively, and of the lines in the plane *ab* which are coplanar with any of the lines of the projective plane not lying in the plane *ab*" (ADG, p. 22).

[35] Whitehead's straight lines are assumed to include an infinite number of points, so that Fano's axiom (see n. 29) has in effect already been introduced.

There is something of a gap, however, between Whitehead's three-dimensional flat "spaces" (some of which presumably have one *time-like* dimension) and the three-dimensional manifold of sense-perception and of physics with its three *space-like* dimensions. The most reasonable assumption, I believe, is that the gap must be filled by something like Whitehead's earlier formulation of extensive abstraction, in particular, the distinction of space from time by means of sense-perception and the derivation of spatial order from temporal order. The technical details of this derivation do not concern us here; it suffices to assume that a manifold with three space-like dimensions is available and that we then lay down axioms of projective order for the points on each straight line in the manifold. At this point the concepts of congruence and distance can be introduced using the methods of von Staudt, Cayley, and Klein described in the previous section. Finally, in our present cosmic epoch:

. . . the ambiguity as to the relative importance of competing families of straight lines (if there be such competing families), and the ambiguity as to the relative importance of competing definitions of congruence, are determined in favour of one family and one . . . congruence-definition. This determination is effected by an additional set of physical relationships throughout the society [PR, p. 149].

Thus Whitehead continues to maintain his distinction between geometry as the science of the necessary (and therefore uniform) relatedness of nature and physics as the science of the contingent (and therefore incompletely uniform) relations of nature. The implications of this distinction are now, however, somewhat clearer: geometry concerns itself with the *possible* types of uniformly curved space—with what is common to all of them (projective geometry) and with what is peculiar to each of them (the three definitions of congruence or distance);[36] physics decides among the three metrical geometries—a decision which "is to be found by comparing the rival theories in respect to their power of elucidating observed facts" (PR, p. 503). (Whitehead never pretends to give any systematic account of the criteria of "elucidation," although he does sometimes allude to simplicity of explanation [see, e.g., R, p. v].) It would appear, then, that geometry is based on the general character of nature as revealed in sense-perception (what was called the "significance" of natural relations in the early works on philosophy of science), and that physics is based on particular features of nature as revealed in sense-perception.

[36] This accounts for Whitehead's belief in the fundamental philosophical importance of projective geometry—a belief which he adopted early and never gave up. Cf. Russell's Preface to his *An Essay on the Foundations of Geometry:* "To Mr Whitehead I owe . . . the inestimable assistance of constant criticism and suggestion throughout the course of construction, especially as regards the philosophical importance of projective Geometry." Russell's own view (in 1897) of the philosophical importance of projective geometry—based on the alleged a priori character of its axioms—is rather different from Whitehead's view. Again, unlike Whitehead, Russell finds nothing philosophically important in the reduction of metrical to projective properties: "This reduction depends, however, except where the space-constant is negative, upon imaginary figures—in Euclid, the circular points at infinity; it is moreover purely symbolic and analytical, and must be regarded as philosophically irrelevant" (p. 28).

What, now, are the legitimate demands, in connection with the subject matter of geometry, on a metaphysical system like Whitehead's philosophy of organism? The minimum demand would be that the system give some indication of those general metaphysical features of the universe which ground the *possibility* of uniform space. A further, less obviously legitimate demand would be that the system explain the general character of the perceptual process by which an organism discovers the regnant form of geometry in its environment. Whitehead attempts to satisfy both demands.

In the first place, the general metaphysical features of the universe which are expressed by the assumptions about extensive connection and regions suffice, according to Whitehead, to prove that "characteristic properties of straight lines, planes, and three-dimensional flat spaces are discoverable in the extensive continuum without any recourse to measurement" (PR, p. 467). Though valid, this is a very weak claim indeed, for the "characteristic properties" in question are limited to those possessed by the straight lines of topology and projective geometry. It is true that the "characteristic properties" of *metrical* straightness and flatness may be defined in projective geometrical terms, but at this crucial point Whitehead deserts his metaphysical analysis and merely alludes to the results of von Staudt, Cayley, and Klein. What we have a right to expect, I should say, is some indication of the metaphysical presuppositions involved in the reduction of metrical to projective geometry—are we, for example, to consider the theory of congruence-groups and the theory of coordinate-transformations as metaphysically fundamental as the theory of extension? Otherwise stated, we have a right to expect from Whitehead an analysis of the circumstances in which the extensive continuum will contain just a single ovate class of regions and also of the circumstances in which all the members of that single ovate class will correspond to convex regions in one of the three usual mathematical senses (Euclidean, hyperbolic, and elliptic).

Perhaps, however, Whitehead's idea is that there is some mode of perception which enables an organism to distinguish one kind of ovateness from another and thereby one kind of uniformly curved space from another. (It is important to note that from Whitehead's point of view the mode of perception in question need not be ordinary sense-perception as we know it and the organisms need not be ordinary living things as we know them.) This hypothesis as to Whitehead's own views about ovateness and flatness seems to be supported by his doubts concerning the number of ovate classes possible in the extensive continuum—doubts which only make sense if one has in mind some means (such as a particular mode of perception) for drastically reducing the infinite variety of logically possible meanings for ovateness and flatness. In Whitehead's early works on natural philosophy it is the mode of perception known as "cognizance by relatedness only" which assures us of the uniformity of space. On the other hand, in his philosophy of organism Whitehead claims he has *explained* certain features of ordinary sense-perception in terms of his previously defined concepts of straightness and flatness. But for this explanation to succeed, the concepts of straightness and flatness employed obviously must be metrical in character (since Whitehead would surely insist that sense-perception conveys *some* feeling, however uncertain and vague, for *distance*).

We are confronted once more, I believe, with the dilemma which came up in Whitehead's reconstruction of kinematics, namely, the apparent impossibility of reconciling abstract mathematical definitions which may be interpreted concretely in many different and incompatible ways, with alleged explanations of the actual data of sense-perception and the actual concepts used in science. Specifically, we are forced to ask the seemingly unanswerable question: how can the *topological* and *projective* properties of Whitehead's flat geometrical elements serve adequately to account for the general *metrical* features of sense-perception (e.g., the recognition of an object moving in approximately a "straight line")?

We turn now to a more detailed discussion of the relation between flatness and sense-perception in the philosophy of organism.

5. Straightness, Flatness, and Presentational Immediacy

Whitehead's conception of two pure modes of perception—"causal efficacy" and "presentational immediacy"—and the doctrine of "symbolic reference" based on the interaction of these two modes occupy an extremely important place in his later writings, especially PR. Here we must necessarily restrict our discussion to the aspects of this theory of perception which bear most directly on the previous analysis of straightness. *Causal efficacy* refers to the awareness of causal objectification, i.e., the felt inheritance of "vector feeling-tone" by an actual occasion from actual occasions in its past. The term "feeling-tone" refers to the fact that the primary content of such inheritance is vague emotion or visceral feeling; the term "vector" refers to the directed character of the inheritance, namely, the feeling of the efficacy of the past occasions in determining the present occasion. It is assumed that in the present cosmic epoch there is an upper limit to the speed of transmission of such vector feeling-tone. Memory and visceral feelings are examples of perception in this mode.

Presentational immediacy is a secondary mode of perception in the sense that it is always derived from perception in the mode of causal efficacy,[37] whereas causal efficacy probably sometimes occurs alone in the case of low-grade organisms (but see below, n. 42). This is the perceptive mode "in which there is clear,

---

[37] The derivation occurs as follows: "A certain state of geometrical strain in the body, and a certain qualitative physiological excitement in the cells of the body, govern the whole process of presentational immediacy. In sense-perception the whole function of antecedent occurrences outside the body is merely to excite these strains and physiological excitements within the body. . . . The perceptions are functions of the bodily states. The geometrical details of the projected sense-perception depend on the geometrical strains in the body, the qualitative sensa depend on the physiological excitements of the requisite cells in the body" (PR, p. 193). Cf. the more elaborate account in PR, pp. 482–83.

distinct consciousness of the 'extensive' relations of the world" (PR, p. 95). Whitehead explains this mode of perception as follows:

Perception which merely, by means of a sensum, rescues from vagueness a contemporary spatial region, in respect to its spatial shape and its spatial perspective from the percipient, will be called 'perception in the mode of presentational immediacy' [PR, p. 185].

Presentational immediacy illustrates the contemporary world in respect to its potentiality for extensive subdivision into atomic actualities and in respect to the scheme of perspective relationships which thereby eventuates. But it gives no information as to the actual atomization of this contemporary 'real potentiality' [PR, p. 188].

Examples of pure presentational immediacy occur in so-called delusive perceptions, e.g., the image of a grey stone in a mirror perceived as illustrating the space behind the mirror, or the feelings in an amputated limb felt as illustrating the space beyond the body. On the other hand, the ordinary perception of a grey stone is a case of symbolic reference in which the word "stone" refers to vaguely localized feelings of efficacy in the immediate past plus similar anticipations for the immediate future, while "grey" refers to the present sense-datum, definite, limited, and sharply located.[38] In general, in *symbolic reference*, feelings derived from causal efficacy are "precipitated" or "projected" upon a region defined by presentational immediacy.

The importance of the two pure modes of perception for our discussion arises from the fact that, given an actual occasion M, seven loci of actual occasions can be defined with the help of these two modes (see Figure 22).[39] First of all, there are three loci definable directly in terms of causal efficacy: (1) *The causal past of M* is that set of occasions such that from each of them there is a "vector flow of feeling" to M. (2) *The causal future of M* is that set of occasions such that to each of them there is a "vector flow of feeling" from M. (3) *The contemporaries of M* form that set of occasions which are neither in the causal past nor in the causal future of M. (The relation of "being contemporary with" is *formally* identical with the relation called "co-presence" which we encountered in the special theory of relativity—see above, p. 19.) Two contemporaries of M need not themselves be contemporary. This suggests another type of locus. (4) *A duration* is a set of mutually contemporary occasions which is "complete" in the sense that any occasion not included in the set is in the causal past or causal future of *some* member of the set. The occasion M may, according to modern relativistic views,[40] belong to an infinite number of durations; in other words,

[38] For Whitehead's detailed discussion of this example, see PR, pp. 184–86.

[39] See PR, pp. 190–92, 486–90. The easiest way of visualizing these seven loci is with the aid of a Minkowski space-time diagram such as Figure 22. In drawing conclusions from such a diagram one must remember that actual occasions are not dimensionless points and that durations are not necessarily Euclidean in character.

[40] Whitehead accepts these views but only on the basis of "scientific examination of our cosmic epoch, and not on any more general metaphysical principle" (PR, p. 191).

the contemporaries of M may be distributed exhaustively (though not exclusively) into a set of durations, each of them including M. Considering any one of these durations including M, we may define two new loci. (5) *The past of a duration* is the set of occasions each of them lying in the causal past of *some* member of the duration. (6) *The future of a duration* is the set of occasions each of them lying in the causal future of *some* member of the duration. (Thus, in Figure 22 the future of the duration d is represented by the entire region above d, and the past of d is represented by the entire region below d.) The seventh locus,

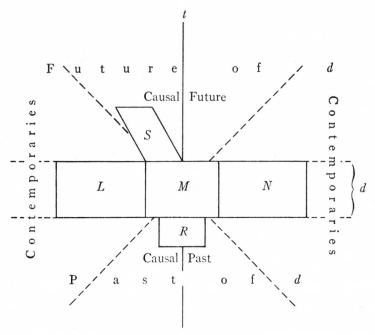

FIG. 22.—Loci of actual occasions

unlike the other six, is not defined in causal terms. (7) *The presented duration of M (if it exists)* is that unique duration indirectly associated with M by means of perception in the mode of presentational immediacy.[41]

The possibility of this seventh locus explains how an occasion (whether "high-grade" like that of a human percipient or "low-grade" like that of a molecule) can single out one duration in which it is "stationary" or "cogredient" and thereby define for itself a particular spatio-temporal system.[42] However,

[41] See PR, p. 257: "The presented locus is a common ground for . . . symbolic reference, because it is directly and distinctly perceived in presentational immediacy, and is indistinctly and indirectly perceived in causal efficacy."

[42] ". . . 'strain-loci' occur as essential components for perception in the mode of presentational immediacy. In this mode of perception there is a unique strain-locus for each such experient. Rest and motion are definable by reference to real strain-loci, and to potential strain-loci. Thus the molecules, forming material bodies for which the science of dynamics is im-

Whitehead is no longer content as in his earlier works merely to *assume* the possibility of a direct intuition of the duration in which an occasion is at rest; instead, the presented duration of any occasion is defined by its close association—amounting, in the case of human percipients in our present cosmic epoch, to practical identity—with a "strain-locus." The latter is defined in terms of straight lines and other flat geometrical elements. Roughly speaking, a *strain* is a feeling (e.g., and typically for human beings, a feeling corresponding to a state of the brain) dominated by the awareness of straight and flat geometrical elements; while a *strain-locus* results from the coalescence of the various strains in an occasion and their *projection* into a single unified locus in the contemporary world. This projection is effected by means of two sets of geometrical elements: (1) the set of points within the volume which constitutes the regional standpoint of the experient occasion, and (2) the set of straight lines defined by pairs of these points. The set of points is the *seat* of the strain; the straight lines are the *projectors;* and the complete region penetrated by the projectors is the strain-locus. The boundaries of a strain-locus will be non-intersecting three-dimensional surfaces. Whitehead remarks that "the theory of 'projection,' . . . requires that the definition of a complete straight line be logically prior to the particular actualities in the extensive environment. This requisite has been supplied by the preceding theory of straight lines" (PR, p. 494). (As already noted, there is considerable doubt as to whether the non-metrical straight lines of Whitehead's theory of extension can serve meaningfully to define what Whitehead calls the "projective" properties of sense-perception.) Since strain-loci are defined by feelings of a certain type in a percipient occasion M, while durations are defined in causal terms, it is not logically necessary that there be *any* duration including M which is closely associated with M's strain-locus. From this Whitehead concludes:

It is an empirical fact that mankind invariably conceives the presented world as consisting of such a [presented] duration. This is the contemporary world as immediately perceived by the senses. But close association does not necessarily involve unqualified identification. It is permissible, in framing a cosmology to accord with scientific theory, to assume that the associated pair, strain-locus and presented duration, do not involve one and the same extensive region. From the point of view of conscious perception, the divergence may be negligible, though important for scientific theory [PR, pp. 492–93].

Three comments may be made about the doctrines just considered. First, we note that Whitehead's account of presentational immediacy is intended as a

---

portant, may be presumed to have unique strain-loci associated with their prehensions" (PR, p. 492). Does this imply that even molecules perceive in the mode of presentational immediacy? One might be tempted to think so, although it is not clear from the passage cited that Whitehead rules out the possibility of strain-loci without presentational immediacy. On the other hand, Whitehead sometimes restricts presentational immediacy to high-grade organisms: ". . . presentational immediacy is an important factor in the experience of only a few high-grade organisms, . . . for the others it is embryonic or entirely negligible" (S, p. 27).

metaphysical or ontological explanation of his earlier views about perception (especially the concepts of significance, durations, and cogredience). Thus, the incorporation of Whitehead's earlier analysis of science into the speculative scheme of his philosophy of organism has led to certain changes in the fundamental concepts of natural philosophy, such as the new distinction between strain-loci and presented durations. And these changes presuppose an innovation in the method of extensive abstraction, namely, the definition of straightness in non-metrical terms.

Second, Whitehead's definition of a duration as a set of contemporary (i.e., causally independent) actual occasions has profound ontological implications. We know that the very possibility of contemporary actual occasions presupposes an upper limit to the velocity of transmission of vector feeling-tone (the basis of all causal influence). It follows that, were durations to be infinitely divisible into actual occasions, one could always find two occasions within a single duration which were causally connectible, thereby violating the definition of a duration.[43] Whitehead avoids this consequence by his cell-theory of actuality,[44] according to which the perpetual becomingness of the universe is a discrete process, so that a finite, non-vanishing interval of time is required for the creation of any definite portion of reality (i.e., of any actual occasion). Whitehead emphasizes the fact that an actual occasion comes into existence *as a whole;* what one refers to as the "earlier" phases of the occasion are not fully determinate and therefore literally non-existent until the occasion completes itself.[45] Thus the actual occasions which constitute a single duration, requiring as they do the same time-span in which to complete themselves, *can* be causally independent of one another. Whitehead uses the phrase "unison of becoming" to refer to this aspect of a duration.

Finally, the categoreal scheme of the philosophy of organism, especially in its emphasis on the fundamental contrast between actuality and potentiality, reinforces Whitehead's earlier distinction between geometry and physics, together with its consequence, the necessity of uniform space-time:

A strain-locus is defined by the "projectors" which penetrate any one finite region within it. Such a locus is a systematic whole, independently of the actualities which may atomize it. In this it is to be distinguished from a 'duration' which does depend on its *physical content.* A strain-locus depends merely upon its *geometrical content.* This geometrical content is expressed by any adequate set of 'axioms' from which the systematic inter-connections of its included straight lines and points can be deduced.

[43] This can easily be seen from a Minkowski space-time diagram, provided one remembers that both durations and actual occasions possess a non-vanishing temporal breadth.

[44] Incidentally, Whitehead derives further support for his cell-theory of actuality from his interpretation of Zeno's paradoxes (see PR, pp. 105–8).

[45] Cf. PR, p. 107: "... in every act of becoming there is the becoming of something with temporal extension; but ... the act itself is not extensive, in the sense that it is divisible into earlier and later acts of becoming which correspond to the extensive divisibility of what has become."

This conclusion requires the systematic uniformity of the geometry of a strain-locus, but refers to further empirical observation for the discovery of the particular character of this uniform system [PR, p. 503].

By way of anticipation it may be noted here that Whitehead is not content merely to urge upon us the necessity of uniform space-time—he actually creates a new theory of relativity, alternative to Einstein's general theory of relativity, formulated in terms of the uniform space-time framework of special relativity. This theory of Whitehead's will be discussed in chapter ix.

6. *Causation and Regions*

In the context of the theory of extension, regions were assumed to be prehensible either through the direct prehension of the actual extensive relationships of actual occasions or through the indirect prehension of potential extensive relationships. This way of analyzing regions—in terms of their prehensible aspects—is appropriate to the theory of extension, which is concerned to show how all fundamental extensive concepts are definable by means of prehensible data. On the other hand, an alternative way of analyzing regions should clearly be both possible and important, namely, by a study of the internal nature of regions construed as loci of interacting actual occasions. Furthermore, such an analysis of regions should be particularly valuable as the basis for a general ontological study of the grouping of actual occasions to form nexūs of various sorts. It is indeed within the context of such an ontological inquiry that Whitehead undertakes to define regions in causal terms. The alternative modes in which regions may be considered are characterized in the following passage (the third mode I interpret as simply the actual perception of a region by one of its component occasions, which furnishes the concrete basis for the two ways of analyzing regions):

Each one of these regions, with its dominant set of ordering relations, can either be considered from the point of view of the mutual relations of its parts to each other, or it can be considered from the point of view of its impact, as a unity, upon the experience of an external percipient. There is yet a third mode of consideration which combines the other two. The percipient may be an occasion within the region, and may yet grasp the region as one, including the percipient itself as a member of it [AI, p. 257].

Regions constitute the simplest, most superficial type of nexus of actual occasions, definable by mere contiguity (spatial or temporal).[46] In order to define the concept of a region we begin by recalling our definition of contemporary actual occasions, namely, two actual occasions are *contemporary* when they are causally independent of each other, i.e., neither is causally objectified

---

[46] Other types of nexus include "Societies, Persons, Enduring Objects, Corporal [*sic*] Substances, Living Organisms, Events, with other analogous terms for the various shades of complexity of which Nature is capable" (AI, p. 255).

in the other (see above, p. 114). Two occasions are *contiguous in time* when (1) they are not contemporary, and (2) there is no occasion which is in the causal past of one of them and in the causal future of the other.[47] In Figure 22 the two occasions M and S are contiguous in time, as are M and R, while R and S are not.

Spatial contiguity is defined in terms of temporal contiguity, as follows. Two actual occasions are *contiguous in space* when (1) M and N are contemporary, and (2) there is no occasion (a) contemporary with both M and N, and (b) such that its causal past includes all occasions, each belonging both to the causal past of M and to the causal past of N. The point of this definition can best be explained by a diagram. In Figure 22 the occasions M and N are contiguous in space, as are L and M, while L and N are not (because M is contemporary with L and N and M's causal past includes all occasions common to the causal pasts of L and N). Contiguity in time and contiguity in space are, of course, incompatible.

A *region* may now be defined as a nexus of actual occasions satisfying certain conditions of contiguity (AI, p. 260).[48] Whitehead does not bother to specify these conditions, but the crucial condition seems obvious, namely, any occasion which belongs to a certain region must be contiguous, either temporally or spatially, with at least one other occasion which belongs to that region. Also, Whitehead might wish to exclude regions with "holes" as he excludes events with "holes" in his earlier works; but this is by no means certain since his later theory of extension is deliberately more general than his earlier theory.

[47] Cf. AI, p. 259. I have paraphrased Whitehead's definition to make evident the role of causation.

[48] The chain of definitions which we are here considering is precisely reversed in the context of the theory of extension; instead of defining regions in terms of contiguity, contiguity is there defined in terms of regions: "Let two actual occasions be termed 'contiguous' when the regions constituting their 'standpoints' are externally connected" (PR, p. 468).

CHAPTER VII

THE THEORY

OF OBJECTS

1. Ingression and Situation of Objects

Having completed a study of the uniform extensional structure of the universe as a whole and of the uniformly curved, four-dimensional spatio-temporal manifold of our particular cosmic epoch, we may turn to the "things" or "objects" which exist, move, and interact in space-time. Here Whitehead is especially concerned to formulate precise definitions for the location, motion, and quantitative properties of "material objects." These definitions may then serve as the means for expressing in mathematical form the laws of motion and interaction of material objects which constitute his theory of relativity.

Objects, it will be recalled, are permanent aspects of nature which recur and hence may be recognized. Recognition, in this usage, is "the non-intellectual relation of sense-awareness which connects the mind with a factor of nature without passage" (CN, p. 143). The *theory* of objects,[1] on the other hand, obviously requires conscious, intellectual comparison of events to discover their similarities and diversities. Objects and events are correlative natural elements; each presupposes the other, since events are the ultimately actual elements in nature while the character of any event is exhausted by the objects which are ingredient in it.[2] "Nature is such that there can be no events and no objects without the ingression of objects into events" (CN, p. 144). Thus *ingression* is Whitehead's term for the general relation of objects to events.[3] The

[1] Throughout this chapter I follow mainly Part IV, "The Theory of Objects," of PNK.

[2] ". . . the character of an event is nothing but the objects which are ingredient in it and the ways in which those objects make their ingression into the event" (CN, pp. 143–44). The "nothing but" in this sentence should probably be interpreted more as a goal for scientific inquiry than as an ontological assertion. As far as natural philosophy is concerned, events are *apprehended* as a unique type of factor in nature. As for ontology, it must be remembered that events are only the most concrete factors *in nature*, and nature is itself an abstraction from the actual universe.

[3] At least, this is the meaning of "ingression" in CN, but Whitehead seems to waver somewhat in his usage. The term does not occur at all in PNK. In R, "ingression" denotes "the complex relationship of those abstract elements of the world such as sense-objects . . . to those other more concrete elements (events) . . ." (p. 37); but "Chairs, tables, and perceptual ob-

various types of objects have varied modes of ingression into events, while even a single type of object may ingress into different events in different modes, and ingression is susceptible of degrees. One particular mode of ingression is especially important for natural science, namely, situation, in which ingression takes on, so to speak, a more "concentrated" form.[4] Whitehead's examples of situation include: an electron with a definite shape and position in space, a gale in mid-Atlantic with a definite latitude and longitude, the cook in the kitchen, the image of an object seen in a mirror (CN, pp. 146–47). The term "situation" is used both for this particular relation of an object to an event and also for the event in which the object is situated. *Situation* "is logically indefinable being one of the ultimate data of science" (PNK, p. 165); nor does Whitehead formulate any general axioms for the relation of situation. However, Whitehead later imposes the following restriction on the possible situations of material objects: two material objects (without common parts) which coexist at the same moment cannot be situated in intersecting events (see below, p. 154, for a discussion of this condition). The situation of an object must be distinguished from its "location" in space, which eventually is to be defined in terms of situation.

The situations of three types of objects are prominent in natural philosophy: sense-objects, perceptual objects, and scientific objects. These three types of

---

jects generally, have lost the complexity of ingression . . ." (p. 38). This suggests that perhaps only sense-objects ingress—or, at least, only sense-objects ingress in the full sense—into events. In UC only sense-objects are explicitly characterized as ingressing into events. Finally, in SMW, Whitehead says that "each eternal object has its own proper connection with each such [actual] occasion, which I term its mode of ingression into that occasion" (p. 229)—but it must be remembered that in SMW neither scientific objects nor physical objects nor perceptual objects are classified as eternal objects.

[4] It must be emphasized, however, that no object is ever, strictly considered, in a single place at a single time: "Finally therefore we are driven to admit that each object is in some sense ingredient throughout nature; though its ingression may be quantitatively irrelevant in the expression of our individual experiences" (CN, p. 145).

The concept of the "physical field" is derivative from this complex ingression of objects in events: "The total assemblage of the modifications of the characters of events due to the existence of an object in a stream of situations is what I call the 'physical field' due to the object. But the object cannot really be separated from its field. The object is in fact nothing else than the systematically adjusted set of modifications of the field. The conventional limitation of the object to the focal stream of events in which it is said to be 'situated' is convenient for some purposes, but it obscures the ultimate fact of nature. From this point of view the antithesis between action at a distance and action by transmission is meaningless" (CN, p. 190). The physical field will be discussed in more detail in connection with Whitehead's theory of relativity (see below, pp. 193–97).

The "conventional limitation" of objects to the events in which they are "situated" is sometimes referred to in discussion of Whitehead's philosophy as "the fallacy of simple location." Now, in the first place, Whitehead himself never calls simple location a fallacy; but, more important, Whitehead contends that simply-located objects (of the physical and scientific types) are essential for natural science, and in fact he formulates a precise definition for the location of such objects (see below in this chapter). The real error—an instance of the fallacy of misplaced concreteness—is to misconstrue these simply-located objects as the concrete elements in nature (cf. SMW, pp. 72 ff., 84 ff.).

objects[5] form an ascending hierarchy, in which each type presupposes the type below it, but no other type of object. Sense-objects, then, are the simplest and most fundamental of all natural objects. Examples of sense-objects are: a specific shade of blue, a specific musical note, a specific odor, etc. The ingression of sense-objects into nature always involves four sets of events (not necessarily distinct): (1) a percipient event, (2) a situation, (3) active conditioning events (which may be more or less distinguishable into generating and transmitting events), and (4) passive conditioning events. Ingression of sense-objects is thus a polyadic relation and hence considerably more complex than the binary substance-attribute relation of much traditional philosophy. As an example, consider the perception of a definite shade of blue in a mirror by a single observer. Then, the percipient event is the relevant bodily state of the observer. The situation is that event in (or better, "behind") the mirror in which the blue is seen. The active conditioning events are the thing which is blue in color, the mirror, and the state of illumination of the room. The passive conditioning events are all the remaining events in nature (which supply the general spatio-temporal framework within which this particular ingression of blueness into nature must be located). This example makes it clear that the situation of an object may be a passive conditioning event; the fact that in "normal" perception the situation of a sense-object is an active conditioning event is what makes natural knowledge possible. But such a generating event of one sense-object generally turns out to be the generating event for a great variety of sense-objects, both for the original percipient event and for many other percipient events. In other words, it is a law of nature that associations of sense-objects in the same situation occur quite insistently in our experience. Such associations of sense-objects in the same situation constitute what Whitehead calls *perceptual objects*. The process of association is a complex one; it involves "(i) the primary recognition of one or more sense-objects in the same situation, (ii) the conveyance of other sense-objects by these primary recognitions, and (iii) the perceptual judgment as to the character of the perceptual object which in its turn influences the character of the sense-objects conveyed" (PNK, p. 89). Taken together, (i), (ii), and (iii) are referred to as a *completed recognition* of a perceptual object.

It must be emphasized once again that the association which issues in perception of perceptual objects is *not* an intellectual affair; rather, the perceptual object is "a factor of nature directly posited in sense-awareness" (CN, p. 155). Whitehead goes out of his way in Note III to the second edition of PNK to repudiate the "class-theory" of perceptual objects, which theory might plausibly follow from the view that a perceptual object is the product of a conscious judgment. However, even in the text of PNK we are told that "the object is more than the logical group; it is the recognisable permanent character

---

[5] In Note III to the second edition of PNK, Whitehead adds a fourth member to the hierarchy, namely, physical objects, which are no longer (as in the texts of PNK and CN) identified with non-delusive perceptual objects.

of its various situations" (PNK, p. 91). Perceptual objects may be delusive or non-delusive. In the former case, the situation of the perceptual object is a passive conditioning event with respect to the ingression of that object into nature. In the latter case, the situation of the perceptual object is an active conditioning event with respect to the ingression of any of its component sense-objects, and this same event can be the situation of the perceptual object for an indefinite number of possible percipient events. Here the situation of the perceptual object is a generating event or "cause." Non-delusive perceptual objects are referred to as physical objects in the texts of PNK and CN, but in Note III to the second edition of PNK Whitehead proposes to distinguish two different types of objects by the qualifying terms "perceptual" and "physical"; an object of the latter type is said to be "a social entity resulting from scientific objects, and halfway towards a perceptual object" (PNK, p. 203). Whitehead's discussion of this proposed distinction is both brief and rather obscure, but his meaning seems to be that a perceptual object is a directly and immediately perceived causal character (of some event) while a physical object is a causal character (of some event) which involves "further discrimination of the significance of immediate appearance" (PNK, p. 204).

Physical objects are of extreme importance in natural science because they are essential to the determination of the generating and transmitting conditions of sense-objects. Thus the situation of a mirror is both the generating condition for that particular association of sense-objects recognized as the physical object called a "mirror" and also the transmitting condition for the sense-objects perceived as images behind the mirror. Similarly, a prism is a physical object whose situation is a transmitting condition for the sense-object which is the spectrum. The trouble with physical objects is that they are too vague to serve as fundamental entities in natural science. The vagueness which afflicts physical objects arises from two sources. In the first place, the object is confused with the sequence of events which are its successive situations. Instead, therefore, of being conceived as a more or less precisely definable abstraction, the object is identified with the inexhaustiveness properly ascribable only to events. Second, even after realizing the abstractness and hence the definability, in principle, of physical objects, it is impossible in practice to define any given physical object in a way which is both precise and also consistent with the group of sense-objects habitually associated with that particular physical object. This last difficulty becomes clear enough if we simply remind ourselves of the puzzle about the oft-darned sock (when does it cease to be the same object?). Another way of putting this difficulty is to say that there is no method analogous to the method of extensive abstraction which is applicable to physical objects.

It is for the purpose of replacing the vague physical objects of common sense as means for characterizing the events active in the ingression of sense-objects into nature that scientific objects have been introduced. Scientific objects (such as atoms and electrons) are not only precise, their relations to one another are also of a highly simple and uniform kind. Now both ordinary physical objects and many scientific objects are of the sort called "material." This suggests the next stage in Whitehead's analysis of natural knowledge: an attempt to develop a precise and systematic account of material objects. It should come as no surprise that, following his reinterpretation of the traditional conceptions of space and time, Whitehead should proceed to reinterpret the third member of the classical "materialistic trinity": matter.

2. Location of Objects; Uniform Objects; Components of Objects

Since the definition of a material object requires reference to various spatial and temporal concepts and since the latter have already been defined by the method of extensive abstraction, we begin by relating the fundamental notion of the theory of objects, situation, to the fundamental abstractive notion, that of an abstractive element. An object is *located* in an abstractive element (i.e., a moment, a volume, an area, a route, or an event-particle, of instantaneous space) if there is a simple abstractive class "converging" to the element and such that each of its members is a situation of the object. Location in instantaneous space is primary, location in time-less space derivative. With respect to a given type of abstractive element there will be various types of location, depending upon the varying relations of objects to parts of their situations. Thus, a definite shade of color located in a given area will also be located in every part of that area, a property that follows directly from the fact that if a color is situated in an event, then it is situated in any part of the event. Precisely this will *not* be true of a physical object like a chair or of a complex sense-object like a tune, the former demanding a definite minimal spatial volume and the latter demanding a definite minimal temporal duration in which to be manifested.

An important distinction emerges from the above examples, that between "uniform" and "non-uniform" objects. In an obvious sense, both the color and the chair are (temporally) uniform, while the tune is not. The notion of uniformity—which will be introduced as one of the defining conditions of material objects—may be precisely defined with the aid of the notion of a "slice" of an event. We define a *slice of an event* e *in a time-system* a (briefly, an *a-slice of* e) as that part of e lying between any two moments of a which intersect e. The two moments are called the *terminal moments* of the slice, and the volumes in which these moments intersect are called the *terminal volumes*. Any a-moment between the terminal moments of an a-slice of e necessarily intersects e in a volume, such a volume being termed an *a-section* of the slice. A slice will always be an event stretching throughout the duration bounded by its terminal moments. The preceding definitions are illustrated by Figure 23.

Uniform objects satisfy the following two laws:

Law I. If a be any time-system and e be a situation of an uniform object O, then an a-slice of e exists which is a situation of O.

Law II. If α be any time-system and *e* be a situation of an uniform object *O* and *e'* be an α-slice of *e* which is a situation of *O*, then every α-slice of *e'* is a situation of *O* [PNK, p. 168].

Law I asserts that a uniform object situated in an event must exist during some finite period of time; while Law II asserts that a uniform object existing during some finite period of time must also exist during any shorter period of time within that period. By defining α-sections of *e* in terms of abstractive classes of α-slices of *e*, we can easily prove that a uniform object *O* situated in *e* is located in every α-section of any α-slice of *e*. This last result represents the capacity of

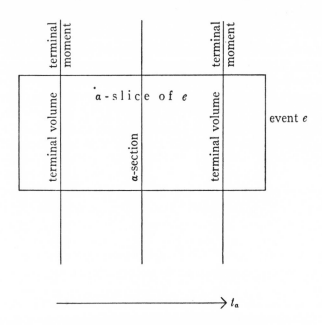

FIG. 23.—Sections and slices of an event *e*

uniform objects for being located in a spatial volume at an instant. All of the preceding characteristics of uniform objects correspond to empirical laws of nature rather than to a priori logical truths.

The notion of a whole and its parts is central in the theory of events and retains its importance in the theory of objects. Since, however, the "whole-part" phraseology has been reserved for events, Whitehead now introduces the terms "main" and "component object." The general notion of the *component of a main object* is, roughly speaking, that of an object which necessarily shares the situation, or some part of the situation, of the main object, but which may also possess a situation independently of the main object. Thus, a certain note may be necessary to a certain tune, but that note can also exist independently of the tune.

Three important types of component objects can now be defined. First of all, a *concurrent component* of an object is a component of the object which, roughly speaking, lasts throughout the same time-interval as the main object. Thus, the color of a certain chair is a concurrent component of the chair, and, equally well, the color of a delusive chair—in a dream, for example—is a concurrent component of the delusive chair. Next, after material objects have been defined, a concurrent component of a main material object, located in some subvolume of that material object, may be defined as an *extensive component* of the main material object. (Incidentally, it will follow from the definition of material objects that an extensive component of a material object is itself a material object.) Thus, the leg of a certain chair is an extensive component of the chair. Finally, the *causal components* of a physical object are defined as "the scientific objects which occupy parts of the situation of the physical object, and whose total assemblage is what constitutes the qualities which are the apparent character which is the physical object apparent in the situation" (PNK, p. 189). Thus, a particular assemblage of electrons, protons, etc., may constitute the set (or a subset) of causal components of a certain chair. Whitehead formulates precise definitions of concurrent and extensive components (see PNK, pp. 169–71), while causal components are presumably not susceptible of precise definition due to the relative vagueness of physical objects.

3. Material Objects

Because of the varied character of material objects it is not possible to define the general concept "material object"; instead, Whitehead defines what it means to compose "a set of material objects of a certain definite sort" (e.g., the class of wooden objects or the class of electric charges). The definition of a class μ of material objects contains five conditions (see PNK, p. 171) which it may be helpful to paraphrase in non-technical language. (1) Every member of μ is uniform, i.e., it can be thought of as located in space at a single instant; (2) only one member of μ can be located in a single place at a single instant; (3) no member of μ can be located in more than one place at a single instant; (4) two members of μ located in non-overlapping places at the same instant are necessarily situated in separated pairs of events; (5) every component, O', of a member, O, of μ is itself a member of μ, which lasts concurrently with O (components such as O' constitute the extensive components of the main object O).

The first three conditions require no comment, since they are simply straightforward formulations of our common-sense beliefs about material objects. The fourth condition may seem logically misplaced inasmuch as it expresses a restriction on the *situations* of material objects in terms of the *locations* of these objects—and location has already been defined in terms of situation. However, the real import of the condition is to restrict the possible extensional relations between the respective situations of a *pair* of material objects, whose respective locations may be assumed to have been already determined; and one can define the location of an object in terms of an abstractive class of its situations *without* inquiring as to the relations of those situations to the situations of some other object. The significance of condition 4 can perhaps be clarified by the following reformulation: If two members of μ, O_1 and O_2, are situated respectively in a pair of intersecting events, then either (a) there is no moment in which both O_1 and O_2 are located, or (b) O_1 and O_2 are located in overlapping volumes (in some moment). Since a pair of situations of O_1 and O_2 respectively could never intersect unless there were *some* common moment in which O_1 and O_2 were located (i.e., some instant at which O_1 and O_2 coexisted), we may ignore alternative (a). Briefly, then, if two material objects of the same sort are to be situated in intersecting events, they must possess a common "part" (more precisely, a common extensive component). Thus, for example, two pieces of wood situated respectively in a pair of intersecting events must have a common part (also a piece of wood) in which they overlap. Since common-sense discourse does not ordinarily refer to the "situation of an object in an event," this fourth condition is not (like the first three conditions) clearly implied by the way we talk about material objects. At least one of Whitehead's own examples of situation—the cook in the kitchen—seems to violate the fourth condition, since the butler could presumably also be situated in the kitchen and yet the cook and butler have no common extensive components. One must conclude that Whitehead was being careless or elliptical in this example of situation; a more accurate statement would be that the cook is situated in a certain part of the kitchen (namely, the part "filled" by the cook's body during some definite span of time).

The example of the cook shows that, according to condition 4, the various situations of a given material object can differ only in their respective "temporal" dimensions (otherwise two material objects of the same sort *could*, without overlapping, be situated in intersecting events). This has serious implications of which Whitehead does not seem to have been fully aware. In particular, it is evident that an abstractive class consisting entirely of situations of a given material object cannot be *simple* because the events which constitute these situations will all have a common spatial boundary (defined by the boundary of the material object itself). Hence Whitehead's definition of location is empty, at least in the case of material objects. To overcome this difficulty one might eliminate from the definition of location the condition that the abstractive class of situation-events be simple—this is Whitehead's course in CN (p. 161), although it is not at all certain that his reason for adopting the new definition had anything to do with his realization of the fault in the old definition. Or, one might reformulate the old definition along the lines of the PNK definition of a stationary prime (see above, p. 70), as follows: An object is located in an abstractive element if there is a simple abstractive class "converging" to the element and such that each of its members extends over some situation of the object.

Condition 5 expresses the common-sense belief that a portion of a material object is itself a material object of the same sort. However, owing to Whitehead's earlier restriction on the topological character of events (and consequently of abstractive elements), it is not now true to say that *any* portion of a material object is itself a material object of the same sort; thus, for example, a piece of wood with a hole in it cannot be an extensive component of some other piece of wood because there are no events which could be situations of a material object with a hole in it. Indeed, we now see that such objects as pieces of wood with holes in them cannot in Whitehead's sense be *located* (since location is defined in terms of situation), and hence presumably do not even qualify as material objects. It is difficult to believe that Whitehead meant to impose such severe limitations on the concept of material objects.[6]

Material objects make possible mathematical physics in its present form, for "It is by means of the properties of material objects that the atomic properties of objects are combined in mathematical calculations with the extensive continuity of events" (PNK, p. 172). Whitehead's example in this connection is the electron, which may be viewed either as an atomic whole or as a continuous spatio-temporal distribution of electricity divisible into extensive components which are similar distributions. It must be emphasized, however, that an electron is more than a mere material object: although the quantitative electric charge of an electron is situated in a series of narrowly circumscribed events (the "occupied" events of the field of that electron) and in this sense constitutes a material object, the electron also ingresses into all the remaining events in nature (the "unoccupied" events of the field of that electron). Furthermore, "the quantitative charge is entirely devoid of character apart from its associated field . . ." (PNK, p. 96). Whitehead ventures to predict that the ultimate scientific objects may eventually turn out to be non-uniform (and hence not "material" even in the limited sense in which electrons and protons are "material"). Such non-uniformity might serve to explain some of the quantum or discontinuous aspects of nature. The development of modern quantum theory beginning in 1924 has confirmed Whitehead's prediction at least to the extent of introducing scientific objects which less and less resemble ordinary material objects. (See below, pp. 214–17.)

To complete the theory of material objects we must give a general account of the motion of such objects and of the extensive quantities which they may possess.

4. The Motion of Material Objects

In order to formulate the theory of the motion of material objects, Whitehead introduces the concept of "stationary events." These turn out to be precisely equivalent to cogredient events,[7] which have already been discussed (see above, pp. 69 ff. and Figure 12); in the interest, therefore, of expository simplicity, I shall omit the definition of stationary events but continue to use the term. Consider an event *e* stationary within a duration *d*. The duration *d* defines

[6] Cf. T. de Laguna's criticism of Whitehead's method of extensive abstraction as insufficiently general, in "Point, Line, and Surface, as Sets of Solids," p. 457, n. 8.

[7] See CN, p. 198: ". . . the 'stationary events' of article 57 of the *Principles* are merely cogredient events got at from an abstract mathematical point of view."

a time-system π. Let the two bounding moments of d be M_1 and M_2 and let a third moment M inhere in d. M will intersect e in a section, the volume V. Now, any other moment of π between M_1 and M_2, say M', will intersect e in a volume V' which is a geometrical replica of V. Also, M_1 and M_2 will intersect e in volumes V_1 and V_2, respectively, which are also geometrical replicas of V and V'. Any volume, such as V, V', V_1, or V_2, in which a moment of π intersects an event cogredient with a duration in π is called a *normal cross-section* of the event (Figure 24, i). An *oblique cross-section* of an event e stationary in π is defined by any moment of another time-system a which intersects e in a volume U but does not intersect either of the terminal volumes of e (Figure 24, ii). All oblique cross-sections formed by moments of a single time-system are geometrical replicas of one another.

Let v_π be the measure of the normal cross-sections of an event e stationary in π, and let v_a be the measure of the oblique cross-sections of e made by the moments of a. On the basis of the transformation equations for space- and time-coordinates in hyperbolic kinematics (equations [5.13]) it is easy to show that v_π and v_a must be related as follows:

$$v_a = \Omega_{a\pi}^{-1} v_\pi , \tag{1}$$

where $\Omega_{a\pi} = (1 - V_{a\pi}^2/c^2)^{-1/2}$, and $V_{a\pi}$ is the velocity of the time-system π in the a-space.

A *normal slice* of a stationary event e can be defined as the slice of e cut off between any two normal cross-sections of e; and an *oblique slice* of e can be defined as the slice of e cut off between any two parallel oblique cross-sections of e (e.g., in Figure 24, ii, the event bounded by U and U' is an oblique slice of e).

We can now define a "motionless" material object in terms of stationary events. A material object is *motionless* within a duration when the object and its extensive components are all situated in events stationary within the duration. For motionless material objects Law I for uniform objects can be restated as follows:

If O be a material object motionless in the duration d and e be the stationary event extending throughout d in which it is situated, then O is situated in any oblique slice of e [PNK, p. 175].

This law is, according to Whitehead, "a fundamental physical law of nature. Namely, percipients cogredient with different time-systems can 'recognize' the same material objects. In other words, the character of a material object is not altered by its motion" (PNK, p. 176). An immediate corollary of this law is that O is located in every oblique cross-section of e.

Uniform motion of a material object is now definable as follows. Let π be the time-system of the duration in which the material object O is motionless; and let a be another time-system in which d' is the duration of maximum extent which intersects e in an oblique slice; then throughout d' in the time-less space of a, O has a uniform motion of translation with the velocity of π in a.

Motions which are not uniform may be *regular*, which means, roughly speaking, that for each point of the (not necessarily rigid) moving material object a time-system can be found in which that point may be considered instantaneously at rest (in other words, at any instant the magnitudes exhibited by each point of the material object will possess values identical with those which would be possessed by the magnitudes exhibited by the corresponding points of a material object at rest). A precise definition of regular motion (which is implicitly presupposed in the usual mathematical analyses of the motion of material objects) can be given in terms of the motionlessness of material objects defined above (see PNK, p. 176).

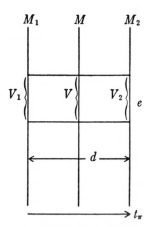

(i) Normal cross-section

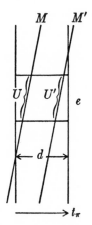

(ii) Oblique cross-section

FIG. 24.—Cross-sections of an event *e*

5. Extensive Magnitudes of Material Objects

We begin with the general notion of a "kind of quantity." To say that two objects O and O' possess quantities of a certain kind and in a definite numerical ratio always presupposes some method of comparison of O and O' which is the defining characteristic of the kind of quantity in question. An *extensive quantity* of a certain kind possessed by a material object O is a quantity which is a determinate function of the quantities of the same kind possessed by any two extensive components of O which (1) exhaust O and (2) are non-overlapping. If the determinate function is simple addition (so that $q = q_1 + q_2$, where q is the quantity possessed by O, and q_1 and q_2 are the quantities possessed respectively by two extensive components of O), then the quantity is *absolutely extensive*. If an extensive quantity is not absolutely extensive it is *semi-extensive*. (Whitehead's example of such a quantity is electromagnetic mass—see PNK, p. 178.) A quantity may be *located* in an abstractive element (see the definition in CN, p. 161). In the case of a very simple material object like an electron, the location of the object itself practically coincides with the location of the quantities possessed by the object (e.g., mass and electric charge). Absolutely extensive quantities may always be thought of as spatial distributions of a quantity of "material" (e.g., mass, electric charge, etc.) throughout a material object. Hence one can always calculate a volume-density, at a time t_a in the a-space of a time-system a, for an absolutely extensive quantity. If we assume that the motion of a material object O is regular, then we can compare the volume-densities ρ_a and ρ_β of an absolutely extensive quantity for the time-systems a and β respectively at a given event-particle P (i.e., at a given "point" in space-time). Let π be the time-system in which the material object O is stationary at P, and let ρ_π be the volume-density at P in π. Then it can be shown that

$$\rho_a \, d \, v_a = \rho_\beta \, d \, v_\beta = \rho_\pi \, d \, v_\pi \, .$$

But from equation (1) we have:

$$d \, v_a = \Omega_{a\pi}^{-1} d \, v_\pi \, , \qquad d \, v_\beta = \Omega_{\beta\pi}^{-1} d \, v_\pi \, .$$

Hence,

$$\rho_a \Omega_{a\pi}^{-1} = \rho_\beta \Omega_{\beta\pi}^{-1} = \rho_\pi \, . \tag{2}$$

Now, taking mutual axes for a and β and making use of the kinematic formulae relating a and β respectively with π, we get from equation (2):

$$\rho_\beta = \rho_a \Omega_{a\beta} \left(1 - \frac{V_{a\beta} \dot{x}_a}{c^2} \right), \tag{3}$$

where $\Omega_{a\beta} = (1 - V_{a\beta}^2/c^2)^{-1/2}$, $V_{a\beta}$ is the velocity of the time-system β in the a-space, and $\dot{x}_a$ is the x_a-component of the velocity $(\dot{x}_a, \dot{y}_a, \dot{z}_a)$ of P relative to π. Letting d/dt_a denote differentiation following the motion of P at $(x_a, y_a,$

z_a, t_a) and letting $\partial/\partial t_a$ denote differentiation at the point (x_a, y_a, z_a), it can be shown that

$$\Omega_{a\pi}\left\{\frac{1}{\rho_a}\frac{d\rho_a}{dt_a}+\mathrm{div}_a\,(\dot{x}_a,\,\dot{y}_a,\,\dot{z}_a)\,\right\}=\Omega_{\beta\pi}\left\{\frac{1}{\rho_\beta}\frac{d\rho_\beta}{dt_\beta}+\mathrm{div}_\beta\,(\dot{x}_\beta,\,\dot{y}_\beta,\,\dot{z}_\beta)\,\right\},\quad(4)$$

where div_a stands for the expression $(\partial\dot{x}_a)/(\partial x_a)+(\partial\dot{y}_a)/(\partial y_a)+(\partial\dot{z}_a)/(\partial z_a)$ and div_β for the expression $(\partial\dot{x}_\beta)/(\partial x_\beta)+(\partial\dot{y}_\beta)/(\partial y_\beta)+(\partial\dot{z}_\beta)/(\partial z_\beta)$. Now, the equation of continuity for an absolutely extensive quantity—which asserts that the quantity is conserved at every point throughout the space of some time-system—can be written as follows:

$$\frac{1}{\rho_a}\frac{d\rho_a}{dt_a}+\mathrm{div}_a\,(\dot{x}_a,\,\dot{y}_a,\,\dot{z}_a)=0\,.\tag{5}$$

Equation (4) shows that if equation (5) holds in any time-system, then it holds in all time-systems (in the terminology of tensor analysis, to be introduced in chapter viii, equation [5] is "Lorentz-covariant").

It is worth pausing at this point to ask exactly what Whitehead's theory of material objects has accomplished. Material objects are defined in terms of their spatio-temporal properties and their special type of situation in events. The motion of material objects—uniform and non-uniform but regular—is defined, again in terms of their spatio-temporal properties and their special type of situation in events. Extensive magnitudes of material objects are defined and shown in special cases to satisfy a differential equation expressing a law of conservation for the magnitude in question. Thus Whitehead ends his treatment of material objects at the point where the usual treatments in theoretical physics begin. But this is what one should expect in view of his concern to lay new, philosophically secure foundations for theoretical physics. Whitehead has demonstrated that his choice of the basic categories of natural philosophy, namely, events, objects, and the relations between them, provides an adequate basis for the standard analysis of the motion and physical magnitudes of material bodies. His remaining task is to show that his natural philosophy also deals satisfactorily with the set of standard philosophical issues that seem to emerge in all attempts to relate perceptual data to the theoretical entities of natural science. These issues center about the meaning of "cause" in natural science and about the interpretation of the "abstracting" or "generalizing" capacity inherent in sense-perception. There is, finally, the question of what Whitehead's natural philosophy implies about the character of living organisms considered as a subject matter for natural science. Stated more precisely, and in Whiteheadian terms, these problems become: (1) In what sense can we say that scientific objects are the causes of physical objects? (2) How are highly generalized objects (such as color and geometrical figure) related to more ordinary sense-objects? (3) Is there anything unique about living objects? To these three questions the next three sections are devoted.

6. Causal Components

For Whitehead the situation of an object in an event and the causal influence of that object on other events are not sharply distinguishable, and both are subsumable under the general notion of ingression. Consequently, there is no special problem of causation in Whitehead's natural philosophy. Whitehead's gradually increasing preoccupation (starting in about 1923) with Hume's problem of the justification of inductive inference is a sign that his natural philosophy is becoming imbedded in a broader metaphysical context. Nevertheless, the fundamental categories of Whitehead's natural philosophy do permit him to reformulate Hume's problem so that it becomes soluble by an appeal to the doctrine of significance. The essence of this reformulation consists in the substitution of the "control-theory" for Hume's (and Russell's) "class-theory" of perceptual objects. Whitehead's argument here involves an appeal to the directly perceived significance of ingressing sense-objects for perceptual objects (see UC, pp. 108–11). When the sense-object is a bodily feeling, the signified perceptual object is recognized with peculiar vividness as some part of the body. (Significance in this "causal" mode is a clear anticipation of the important concept of "causal efficacy" in Whitehead's later metaphysics.) When the sense-object is, so to speak, projected outside the body, the signified perceptual object may be "vague, illusive, or absent"; nevertheless, in favorable circumstances perceptual objects like tables, trees, stones, etc., *are* recognized as controls of ingression. This is possible because "What we observe is the control in action during the specious present" (UC, p. 111).

The principal problem connected with the concept of cause in natural philosophy is, for Whitehead, not the epistemological problem of induction but a purely descriptive problem: how are the objects referred to in science as "causes" logically related to the "wholes" of which they are "parts." Whitehead's discussion of causal components begins with a paradox: on the one hand, physical objects are, roughly speaking, *continuous* in space and time (and therefore belong to the class of material objects); on the other hand, science tells us that physical objects are "composed" of a *discrete* collection of scientific objects (molecules, or atoms, or electrons and protons). To resolve the paradox, we must remember that all objects—whether physical or scientific—are abstract, and therefore limited, aspects of concrete events which are the situations of these objects. Thus, the immediate appearance of a given drop of water is the apparent character of a certain event and therefore a material object; while the causal character of the same event is an aggregate of "atomic" entities (or scientific objects) and therefore not a material object. Each scientific object represents the set of relatively permanent properties of the mutual relations between the drop of water considered as a physical object and other physical objects (in particular, the physical objects which are our scientific instruments). Scientific objects turn out to be characters of characters, second order characters, of a different logical type from apparent characters;[8] hence there is no incompatibility between the concept of a *continuous* physical object situated in an event and the concept of a *discrete* collection of scientific objects situated in that same event.

The history of science may be viewed as a succession of discoveries of progressively simpler, more permanent, and more self-sufficient scientific objects. Whitehead mentions four crucial stages in this history: (1) Archimedes' discovery of specific gravity, (2) Newton's discovery of mass, (3) Dalton's atomic theory, and (4) the concept of the ether, i.e., the concept of events in space empty of appearances. This list clarifies Whitehead's interpretation of scientific objects as certain constant functions of the relations among physical objects. (1) The specific gravity of a substance is a measure of the weight of a fixed volume of the substance as compared with the weight of an equal volume of some other (standard) substance. (2) The mass of a body is a measure of the quantity of "inertia" possessed by the body compared with some other (standard) body. (3) Atomic weights, in Dalton's sense (and also in the sense of modern chemistry), are defined directly as functions of the relative combining weights of the elements. (4) In the case of the ether the relevant scientific objects (say, the energy of the electromagnetic field *in vacuo*) must be defined as complicated functions of physical objects (e.g., electric charges and magnetic poles) in nonempty parts of space.

Another paradox which confuses much discussion of the relation between sense-objects and scientific objects arises from the concrete-abstract contrast. Are sense-data concrete and electrons abstract? Or conversely? For Whitehead, it all depends on one's point of view. Assuming that one wants to continue using the notoriously ambiguous terms "concrete" and "abstract" (which Whitehead prefers to avoid in his later metaphysical works), one's conclusion must be that in the sequence by which scientific knolwedge is derived from sense-experience, sense-objects are more concrete and scientific objects more abstract; while in the sequence from cause to effect, scientific objects are more concrete and sense-objects more abstract.

If we follow the route of the derivation of knowledge from the intellectual analysis of sensible experience, molecules and electrons are the last stage in a series of abstractions. But a fact in nature has nothing to do with the logical derivation of concepts. The concepts represent our abstract intellectual apprehension of certain permanent characters of events, just as our perception of sense-objects is our awareness of qualities of nature resulting from the shifting relations of these characters. Thus scientific objects are the concrete causal characters, though we arrive at them by a route of apprehension which is a process of abstraction. In the same way, what, in the form of

---

[8] Cf. CN, p. 40: "What . . . is it that science is doing . . . ? My answer is that it is determining the character of things known, namely the character of apparent nature. But we may drop the term 'apparent'; for there is but one nature, namely the nature which is before us in perceptual knowledge. The characters which science discerns in nature are subtle characters, not obvious at first sight. They are relations of relations and characters of characters."

a sense-object, is concrete for our awareness, is abstract in its character of a complex of relations between scientific objects. Thus what is concrete as causal is abstract in its derivation from the apparent, and what is concrete as apparent is abstract in its derivation from the causal [PNK, pp. 188–89].

To avoid misunderstanding of such a passage as this, one must remember that for Whitehead *both* sense-objects and scientific objects are abstract when compared with events, the maximally concrete natural elements, *both* in the order of acquiring knowledge and in the order of causation. To ascribe full concreteness *either* to bits of matter (as in the materialistic interpretation of Newtonian physics) *or* to electrons and protons (as in some interpretations of modern physics) is to commit the fallacy of misplaced concreteness.

7. Figures

Whitehead proceeds next to supplement his analysis of the varying degrees of abstractness attributable to an object in different contexts with a discrimination of several types of highly generalized objects called "figures," among which "sense-figures" are primary and "geometrical figures" derivative. Very roughly, a sense-figure is the actually perceived "spatial shape" of an event in which a sense-object is situated. A given sense-object (e.g., a definite shade of blue) will in general possess a whole family of sense-figures, differing from the sense-figures of some other sense-object. Thus, for example, smells and tastes possess barely perceptible sense-figures quite unlike the sense-figures of sights and touches. Figures arise from the perception of the relation of sense-objects to the extensive properties of their situations. Specifically, if in a certain time-system the instantaneous volumes of the situation of some sense-object are all practically congruent, it becomes possible for an observer in that time-system to perceive a sense-figure. Of course, in order for the sense-figure actually to be perceived, the situation of the sense-object must be sufficiently prolonged. This will be the case when the situation of the sense-object and some percipient event are both cogredient within the same duration. A precise definition may be formulated as follows: The *sense-figure*, for a time-system a, of a sense-object O in a situation σ is that unique relation which holds between any a-volumes congruent with one of the a-sections of σ.

Sense-objects, and thereby sense-figures, may be generalized in several directions (but the generalization must *not* be thought of as necessarily a temporal process starting from *less* general objects). Suppose one confines attention to sense-objects of a given sense (e.g., sight). Then, the analogy between sight-objects of a particular quality (e.g., blue of all shades) leads to the possibility of a *generalized sight-object* (e.g., blue). Correspondingly, there are generalized sense-figures (e.g., the sense-figure of blue). The ultimate generality in this direction is reached with sense-objects such as color or touch or taste (and perhaps also the corresponding sense-figures, "visual" surface, "tactile" surface, etc.). However, by generalizing, not with respect to sense-objects of a single sense, but rather with respect to sense-objects of different senses, one can obtain objects independent of the peculiarities of any particular sense. We have already encountered one such object, namely, sense-figure itself. A further stage of generalization in this direction leads to the concept of figures which are independent of the senses altogether; these are *geometrical figures*.[9] Our ability to perceive geometrical figures is connected with our awareness of perceptual objects, each of which is itself dependent upon the natural association of a set of sense-objects in a single situation: "The high perceptive capacity of sense-figures leads to their association in a generalised figure, which is the geometrical figure of the object. Indeed, the insistent obviousness of the geometrical figure is one reason for the perception of perceptual objects" (PNK, p. 193). Thus, a more generalized object may very well be easier to recognize than a less generalized object, e.g., a dark-blue figure may be noticed before the dark-blueness, and a circular shape may be noticed before the dark-blue sense-figure.

The preceding considerations help to clarify Whitehead's general conception of "the principles of natural knowledge." When we read about "deduction of scientific concepts from the simplest elements of our perceptual knowledge" (PNK, p. vii) as constituting the program of Whitehead's natural philosophy, we are not to identify what is "simplest" in the direct act of sense-perception with what is "simplest" according to some scheme of categories for analyzing sense-perception. "Perceptive insistency is not ranged in the order of simplicity as determined by a reflective analysis of the elements of our awareness of nature" (PNK, p. 192). If pure sense-qualities were in fact the only directly perceived data, then it would indeed be a mystery as to how the spatial and temporal concepts of natural science could ever be applied to these data. Since, however, geometrical figures are "insistently obvious" in ordinary sense-perception, the mystery is dispelled and the method of extensive abstraction need not appear as a logical tour de force or sleight of hand. Finally, one must avoid confusing Whitehead's interpretation of the derivation of spatial and temporal concepts with the doctrine which attempts to derive space and time from relations among *figures*. This latter view (exemplified by Poincaré's discussion of geometry)[10] leads to an irreducible and therefore incoherent multiplicity of spaces and times: tactile space, visual space, etc.; tactile time, visual time, etc. Whitehead's procedure, on the other hand, by which space and time are derived from uniform relations among *events*, leads to the single spatio-temporal framework presupposed in *both* natural science and daily life.

[9] The distinction between sense-figures and geometrical figures is reminiscent of the Aristotelian distinction between physical and geometrical concepts. Thus, to use one of Aristotle's favorite examples, "snub" is a physical concept since it refers to a certain definite kind of physical material, namely, the flesh of a nose; while "curved" is a geometrical concept since it refers to no definite kind of physical material but rather to a possible shape for the surface of any physical thing.

[10] H. Poincaré, *Science and Hypothesis*, chap. iv.

8. Rhythms

Percipient objects (or individual minds) are somehow beyond nature and hence outside the scope of Whitehead's natural philosophy. However, since living organisms are parts of nature so that biological phenomena are not *essentially* different from physical phenomena (see PNK, p. 3), some account of what distinguishes the living from the non-living should be possible *within* Whitehead's natural philosophy. We must not except to find a class of objects all possessing the property "living"; rather, life consists in a special relation between an object of a certain type and the stream of events which constitute the successive situations of that object. As for the object, it is not uniform but "rhythmic," i.e., requiring definite minimal periods of time in which to complete itself. As for the events associated with life, they are always, as far as we now know, situations of physical objects. But what we have said so far does not really distinguish life from non-life, since (1) many non-uniform objects have no special relation to life, and (2) physical objects are obviously not sufficient conditions of life. (Incidentally, the reconciliation of the uniformity of physical objects with the non-uniformity characteristic of life occurs through the causal components—non-uniform, periodic, scientific objects, e.g., atoms and molecules—of the physical objects.) As already suggested, life must be identified as a special relation between an object and its successive situations—the object being a complex but self-identical pattern, and the successive situations being more or less different from one another and providing a perpetually novel background against which the recurring pattern exhibits itself. This special relation of a pattern and its successive situations may be called a *rhythm*, which thus turns out to be neither a mere event nor a mere object but rather a unique type of natural element.

Rhythms, as Whitehead conceives them, seem to be intermediate in concreteness between objects and events. On the one hand, a rhythm must clearly be less concrete than an event because the rhythm is a function of some recurrent pattern (unlike events, which are unique). On the other hand, "A rhythm is too concrete to be truly an object. It refuses to be disengaged from the event in the form of a true object which would be mere pattern" (PNK, p. 198). A rhythm is, so to speak, an "embodied" or "realized" pattern, not quite the same each time it occurs. The source of this perpetual novelty or spontaneity—which is of the essence of life but which can never be characterized in general terms—is the "passage" inherent in every event. The mention of passage (which becomes "process" or "creative advance" in Whitehead's later works) is an indication that the discussion is encroaching on the metaphysics of the philosophy of organism, in the context of which it is possible to formulate a more detailed and adequate account of life. Consideration of this account will be reserved for chapter x.

CHAPTER VIII

THE LAWS OF MOTION,

GRAVITATION, AND

ELECTROMAGNETISM

IN CLASSICAL AND

RELATIVITY PHYSICS

1. Motion, Gravitation, and Electromagnetism in Classical Physics

Discussion of Whitehead's theory of relativity is best preceded by an exposition of the theory which it is designed to replace, namely, Einstein's general theory of relativity. In fact, as we shall see, it is not quite accurate to speak of Whitehead's theory as "replacing" Einstein's, any more than we should speak of Whitehead's version of the special theory of relativity as "replacing" Einstein's; Whitehead is concerned to *reinterpret* what he takes to be the genuine insights and achievements of the general theory of relativity in terms of the categories of his own natural philosophy. As a background to Einstein's general theory of relativity I shall sketch the classical, pre-relativistic theories of motion, gravitation, and electromagnetism and the special relativistic modifications of these theories.

We begin with the notion of a three-dimensional rectangular Cartesian coordinate-system with respect to which the positions of material particles may be specified by ordered triples of real numbers. Within the infinite set of such coordinate-systems in all possible states of relative motion, there is an infinite subset of coordinate-systems—called "inertial systems"—with respect to which Newton's laws of motion are valid. (In Newtonian mechanics it is, of course, unnecessary to specify any definite time-coordinate, since time is assumed to be absolute, i.e., the same for all possible choices of spatial axes.) We select any one of the inertial systems (call it K) with axes x, y, z. Now let the position of a particle with mass m be (x, y, z) at time t; let the velocity of m at t be $(\dot{x}, \dot{y}, \dot{z})$, where $\dot{x} = dx/dt$, etc.; and let the acceleration of m at t be $(\ddot{x}, \ddot{y}, \ddot{z})$, where

165

$\ddot{x} = d^2x/dt^2$, etc. Finally, let the components of force along the three axes acting on m at t be (f_x, f_y, f_z). Newton's first and second laws of motion may be expressed as follows:

$$f_x = m\ddot{x}, \qquad f_y = m\ddot{y}, \qquad f_z = m\ddot{z}. \tag{1}$$

In words: A material particle acted on by an impressed force is accelerated in the direction of the force, the magnitude of the acceleration being equal to the ratio of the force to the mass of the particle.

Newton's third law of motion concerns the interaction of a pair of material particles. Let the masses of these particles be m_1 and m_2; let the force exerted by particle 2 on particle 1 be $(f_{12,x} f_{12,y} f_{12,z})$ and let the force exerted by particle 1 on particle 2 be $(f_{21,x} f_{21,y} f_{21,z})$. Newton's third law of motion (the principle of action and reaction) may be expressed as follows:

$$f_{12,x} = -f_{21,x}, \qquad f_{12,y} = -f_{21,y}, \qquad f_{12,z} = -f_{21,z}. \tag{2}$$

In words: For every force exerted by one material particle on another, there is an equal but opposite force exerted by the second particle on the first particle.

It will be noted that our verbal formulations of Newton's laws of motion do not involve any explicit reference to coordinate-systems. This feature of the verbal formulations obviously makes for economy of expression and can easily be transferred to the algebraic formulations with the aid of "vectors," which will be expressed by boldface letters. (See Appendix III for a brief account of vectors.) Thus, equations (1) can be replaced by the single vector equation:

$$f = ma, \tag{3}$$

where f and a are parallel vectors and m is a scalar. Analogously, equations (2) can be rewritten:

$$f_{12} = -f_{21}. \tag{4}$$

By appealing to the invariance of mass and the vector character of acceleration and force (in Newtonian mechanics) we can readily demonstrate that if Newton's laws of motion hold in K, they must hold in any other inertial system K'; mathematically, we say that Newton's laws of motion are *covariant* with respect to Galilean transformations (i.e., transformations which carry one inertial system into another). The Galilean transformation equations for two inertial systems, K and K', have the following form:

$$\left.\begin{aligned} x' &= x - v_x t, \\ y' &= y - v_y t, \\ z' &= z - v_z t, \\ t' &= t, \end{aligned}\right\} \tag{5}$$

where K and K' have a common origin at $t = t' = 0$, the corresponding axes of K and K' are parallel to each other, and $v_x, v_y,$ and v_z are the components of the velocity of K' relative to K.

Not included in Newton's explicit formulation of the laws of motion, but definitely accepted by him (and also by most later physicists until the nineteenth century) are several additional restrictions on the force-pairs occurring in the principle of action and reaction, namely, action and reaction are assumed to be directed along the line joining the two particles and to be a function only of the distance between the particles (velocity and acceleration being irrelevant). When these last conditions hold, it is always possible to find a "potential energy function" V, dependent only on the distance between two particles (and assumed to equal zero when the two particles are infinitely far apart), whose negative first derivative represents the force of interaction between the two particles. Algebraically:

$$\left.\begin{aligned} f_{jk,x} &= -\partial V_{jk}(r_{jk})/\partial x_j, \\ f_{jk,y} &= -\partial V_{jk}(r_{jk})/\partial y_j, \\ f_{jk,z} &= -\partial V_{jk}(r_{jk})/\partial z_j, \end{aligned}\right\} \quad (6)$$

where $f_{jk,x}$ is the x-component of the force acting on the jth particle due to the kth particle, V_{jk} is the potential energy function of the jth and kth particles, and x_j is the x-coordinate of the jth particle; and similarly for $f_{jk,y}$ and $f_{jk,z}$. Introducing vector notation, we have:

$$f_{jk} = -[\partial V_{jk}(r_{jk})/\partial r_{jk}]\,\hat{r}_{kj}, \quad (7)$$

where f_{jk} is the force acting on the jth particle due to the kth particle, r_{jk} is the distance between the jth and kth particles, and $\hat{r}_{kj}$ is the unit vector pointing from the kth particle toward the jth particle. When a potential energy function for the interaction of N particles,

$$V = \sum_{j=1}^{N} \sum_{k=j+1}^{N} V_{jk}(r_{jk}), \quad (8)$$

can be found, the forces of interaction are said to be "conservative." Examples of such forces are those of gravitation, electrostatics, and magnetostatics.

Newton's program for "natural philosophy" may be characterized as the attempt to discover all the forces of nature (or "natural powers"), each such force to be defined by a definite mathematical function of the distance between two particles and of certain other properties of the two particles (such as mass or electric charge but exclusive of velocity and acceleration). Given a particular force-function of x, y, z, and t, one obtains from equations (1) a set of three ordinary second order differential equations, which can be solved for x, y, and z as functions of t. Newton discovered the law of gravitational force; later, Coulomb's laws of electrostatic and magnetostatic force turned out to be closely similar to the law of gravitation. However, as we shall see, the Newtonian program in its strict sense foundered on the phenomena of electrodynamics (or electromagnetic forces).

Newton's law of gravitation states: "Any two material particles attract one another with forces directed along the line joining the two particles and whose magnitudes vary directly with the product of the masses of the two particles and inversely as the square of the distance separating the two particles." Algebraically:

$$f_{12} = -f_{21} = G \frac{m_1 m_2}{r_{12}^2} \hat{r}_{12} , \qquad (9)$$

where f_{12} is the force exerted on particle 1 by particle 2, f_{21} is the force exerted on particle 2 by particle 1, m_1 and m_2 are the respective masses of particle 1 and particle 2, r_{12} ($= [(x_1 - x_2)^2 + (y_1 - y_2)^2 + (z_1 - z_2)^2]^{1/2}$) is the distance between particle 1 and particle 2, G is a universal constant, and $\hat{r}_{12}$ is the unit vector pointing from particle 1 toward particle 2. It is obvious that this law is covariant with respect to the Galilean transformation. Also, the law unmistakably involves "action at a distance," since the two particles must be supposed to exert mutual gravitational forces (given by equation [9]) on each other *at the instant* they are placed a distance r_{12} apart.

It can be shown by a study of the properties of the differential equations expressing Newton's laws of motion that, given the initial position and velocity of each particle in a mechanical system and given the forces of interaction among the particles in the system, Newton's laws of motion completely determine the future behavior (as well as the past history) of the system. Thus the *state* of a mechanical system can be defined by the positions and velocities of each particle in the system; the succession of states is then regulated by the internal forces of the system (assuming the system to be isolated from all external forces). In general, to solve any problem involving the motion of a system of bodies under the influence of their mutual gravitational forces, one must compute the total gravitational force on each body due to all the other bodies in the system, and then find the trajectory of each body resulting from its initial state and the impressed gravitational forces acting on it. Although it is easy enough to set up the appropriate differential equations for the general case of n bodies moving under the influence of their mutual gravitational forces, it turns out that, for $n > 2$, explicit closed solutions of the differential equations in terms of known mathematical functions have never been found; consequently one must be content with approximate solutions for various special cases of the n-body problem. These approximations may be indefinitely improved, so that in practice (e.g., in astrophysics) Newtonian gravitational theory is eminently satisfactory (ignoring relativistic effects, of course).

Suppose, now, that one studies the behavior of moving particles with respect to coordinate-systems which are *not* inertial. What one finds, in general, is that new components of acceleration appear for each of the particles, and that these accelerations cannot be accounted for in terms of genuine physical forces—at least, not in terms of such forces as are envisaged in the Newtonian program. Let us consider first the case of a coordinate-system U moving with uniform acceleration a along the negative x-axis of an inertial system K. For simplicity we assume that U and K coincide at $t = 0$ and that U starts from rest at $t = 0$. Our problem is to compare the motion of a particle of mass m as described in K and in U. Let the position of m in K be represented by the vector r and the position of m in U by the vector r'. From the Galilean transformation equations (5) we get immediately:

$$r = r' - a\,\frac{t^2}{2}. \tag{10}$$

Differentiating equation (10) twice with respect to the time gives:

$$\ddot{r} = \ddot{r}' - a\ .$$

Thus for a given impressed force f in K, there will be an "effective" impressed force f' in U given by:

$$f' = f + ma\ . \tag{11}$$

In other words, in addition to f a new force will appear in K; this force, expressed by the quantity ma, is called an *apparent* or *inertial* force. Such a force occurs, for example, on a body moving in a freely falling box: here K may be taken as a coordinate-system fixed to the earth; U is the coordinate-system of the box; and the magnitude of a is equal to g, the acceleration of gravity. It follows that the impressed gravitational force on m (i.e., the weight of m, or $-mg$) is exactly balanced by the inertial force mg, and the particle in the freely falling box appears not to be acted on by any impressed forces at all.

We consider next the case of a coordinate-system R in a state of uniform rotational motion ω with respect to the inertial system K. For simplicity we assume that the origins of R and K coincide and we take the axis of rotation through the common origin. Let the position vector of a particle of mass m be r in both K and R. Then it can be shown that[1]

$$u = u' + \omega \times r\ , \tag{12}$$

where u is the velocity of m with respect to K and u' is the velocity of m with respect to R. It follows from equation (12) that

$$a = a' + 2\,(\omega \times u') + \omega \times (\omega \times r)\ , \tag{13}$$

where a is the acceleration of m with respect to K and a' is the acceleration of m with respect to R. Thus two new components of acceleration have appeared in the inertial system K. Therefore, the effective impressed force in R will be given by:

$$f' = f - 2m\,(\omega \times u') - m\omega \times (\omega \times r)\ , \tag{14}$$

where f is the impressed force on m in K, f' is the impressed force on m in R, $2\,m(\omega \times u')$ is the so-called Coriolis force, and $m\omega \times (\omega \times r)$ is the familiar centrifugal force. If m is rigidly attached to R, then u' vanishes and so does the Coriolis force.

---

[1] For details see H. Goldstein, *Classical Mechanics*, pp. 132 ff., or G. Joos, *Theoretical Physics*, pp. 232 ff.

As in the case of the uniformly accelerated coordinate-system we notice that no bodies can be found in which to locate the sources of centrifugal and Coriolis forces. Furthermore, centrifugal force *increases* with distance from the axis of rotation—a most unusual characteristic for a "real" force. Once more, from the Newtonian point of view, these forces must be termed "fictitious." (Of course, this does not mean that the dynamical theory of such forces is useless; in fact, the theory has many important applications in dealing with rotating bodies, e.g., tops, gyroscopes, planets, and molecules.)

Both parts of Newtonian gravitational theory—the laws of motion as well as the law of gravitation itself—were reformulated later in ways which were at once mathematically more convenient and heuristically more suggestive. Let us consider first the laws of motion. By introducing the concept of energy, it is possible to recast the laws of motion, in an analytically more powerful form, for a large class of mechanically important cases (namely, those in which only conservative forces are involved). The total potential energy V of a system of N particles is defined by the forces of interaction among the particles, i.e., by equations (6) and (8); the total kinetic energy T of a system of N particles is defined as the sum of the kinetic energies of the individual particles:

$$T = \sum_{j=1}^{N} \tfrac{1}{2} m_j v_j^2, \tag{15}$$

where m_j is the mass of the jth particle and v_j is the speed of the jth particle. We may now state "Hamilton's principle": "The path actually traversed by a moving particle between the times t_1 and t_2 is that path for which a certain function L of the total system—defined as $L = T - V$ and called 'the Lagrangian'—has a stationary value (i.e., a minimum or maximum value as compared with its value along any other path with the same end-points)." In mathematical terms:

$$\delta \int_{t_1}^{t_2} L \, dt = 0 \qquad (L = T - V), \tag{16}$$

where t is the time and δ means variation of the integral for infinitesimal differences in path, the end-points (x_1, y_1, z_1, t_1) and (x_2, y_2, z_2, t_2) being kept fixed. (It is possible to formulate a generalized version of Hamilton's principle for non-conservative systems, but this requires the reintroduction of the concept of force.) Equations (8), (15), and (16) are together mathematically equivalent to Newton's laws of motion (and hence, of course, are covariant with respect to the Galilean transformation).

It is convenient to express equation (16) in differential form, that is, to find a set of differential equations which will insure the stationary character of the integral $\int_{t_1}^{t_2} L \, dt$. These differential equations, called the "Euler-Lagrange equations," can be shown to have the form:

$$\frac{\partial L}{\partial x} - \frac{d}{dt}\left(\frac{\partial L}{\partial \dot{x}}\right) = 0 \,,$$

$$\frac{\partial L}{\partial y} - \frac{d}{dt}\left(\frac{\partial L}{\partial \dot{y}}\right) = 0 \,,$$

$$\frac{\partial L}{\partial z} - \frac{d}{dt}\left(\frac{\partial L}{\partial \dot{z}}\right) = 0 \,,$$

(17)

where $\dot{x} = dx/dt$, $\dot{y} = dy/dt$, $\dot{z} = dz/dt$.

Now, the Euler-Lagrange equations are covariant, not only with respect to spatial orthogonal coordinate transformations, but also with respect to general spatial coordinate transformations. This last fact explains the superiority of the Lagrangian laws of motion over the Newtonian laws of motion. If it is possible to express the total potential and kinetic energies of a mechanical system in terms of *any* set of variables, q_i, and their corresponding first time derivatives, $\dot{q}_i$, then—regardless of the physical significance of these variables—it is possible to use the Lagrangian laws of motion for analyzing the behavior of the system. The variables q_i are called "generalized coordinates of position" and the variables $p_i \, (= \partial L/\partial \dot{q}_i)$ "generalized momenta." The Euler-Lagrange equations in terms of generalized coordinates have exactly the same form as before:

$$\frac{\partial L}{\partial q_i} - \frac{d}{dt}\left(\frac{\partial L}{\partial \dot{q}_i}\right) = 0 \qquad (i = 1 \ldots n) . \quad (18)$$

(The Euler-Lagrange equations may be generalized to include non-conservative systems, but this requires the introduction of so-called "generalized forces.") An obvious field for the application of these laws of motion is to the behavior of extended rigid bodies: it is clearly impossible to formulate completely the forces of interaction among all the particles of such a body, and yet it is often quite feasible to express the total potential and kinetic energies of the body in terms of a small number of parameters (e.g., the coordinates of the center of gravity, etc.).

The Lagrangian formulation of mechanics can be extended to the case of coordinate-systems rotating with respect to inertial systems by adding to the expression for the Lagrangian another potential energy term:

$$-\frac{1}{2}\sum_{j=1}^{N} m_j r_j^2 \omega^2 \,,$$

where m_j is the mass of the jth particle, r_j is the radial distance of the jth particle from the axis of rotation of R in K, and ω is the magnitude of the angular velocity of R with respect to K.[2]

We turn now to the reformulation of the law of gravitation. Consider N particles exerting mutual gravitational forces on one another. Letting f_j stand

[2] See E. T. Whittaker, *Analytical Dynamics*, p. 41.

for the total force acting on the jth particle, we have, by the definition of the potential function,

$$f_j = - \sum_{k=1}^{N} \frac{\partial V_{jk}}{\partial r_{jk}} \hat{r}_{kj} \qquad (j \neq k). \quad (19)$$

But by Newton's law of gravitation:

$$f_j = - \sum_{k=1}^{N} G \frac{m_j m_k}{r_{jk}^2} \hat{r}_{kj} \qquad (j \neq k). \quad (20)$$

Therefore:

$$V_{jk} = - G \frac{m_j m_k}{r_{jk}}; \qquad (21)$$

or, considering one of the particles as a test-particle with unit mass, one finds for the gravitational "potential" ψ of a particle of mass m at distance r:

$$\psi = - \frac{Gm}{r}. \qquad (22)$$

We can now express the gravitational potential energy function as:

$$V = - \sum_{j=1}^{N} \sum_{k=j+1}^{N} G \frac{m_j m_k}{r_{jk}}. \qquad (23)$$

Also, for a continuous distribution of matter, equation (22) can be replaced by:

$$\psi = - G \int_m \frac{dm}{r} = - G \int_\tau \frac{\sigma \, d\tau}{r}, \qquad (24)$$

where σ is the mass density and τ is volume. Poisson showed that, in the case of a discrete or continuous distribution of matter, ψ satisfies the second order partial differential equation:

$$\frac{\partial^2 \psi}{\partial x^2} + \frac{\partial^2 \psi}{\partial y^2} + \frac{\partial^2 \psi}{\partial z^2} = 4 \pi G \sigma. \qquad (25)$$

In empty space, that is, when $\sigma = 0$, Poisson's equation becomes:

$$\frac{\partial^2 \psi}{\partial x^2} + \frac{\partial^2 \psi}{\partial y^2} + \frac{\partial^2 \psi}{\partial z^2} = 0, \qquad (26)$$

which is Laplace's equation.

Equation (25) is the classical equation of the gravitational field. Like Newton's law of gravitation it involves the notion of "action at a distance," since it implies that any change in the density of matter in one region will be instantaneously propagated to remote regions. Moreover, even the field theory of gravitation in its classical form presupposes material particles as the fundamental physical reality; the gravitational field is not a self-subsistent entity but merely an adjunct or resultant of the material particles. Thus, a gravitational field is inconceivable in the absence of a *source* (i.e., a piece of matter), while gravitational potential energy manifests itself only in the actual presence of a material particle. The concept of a field in classical gravitational theory is, therefore, little more than the concept of a continuous mathematical function of the three spatial coordinates, x, y, z, representing the force acting on a test-particle at any point (x, y, z).[3]

We come next to the classical electromagnetic theory of Maxwell, Hertz, and Lorentz. To appreciate Maxwell's achievements one must recall the failure of all attempts, during the first half of the nineteenth century, to construct a unified theory of electrical and magnetic phenomena in accordance with the pre-scriptions of Newton's program for natural philosophy (the laws of motion and conservative forces). Some theories of electricity and magnetism abandoned the principle of action and reaction, others introduced non-conservative forces, until, finally, Maxwell sidestepped, albeit somewhat reluctantly, the whole Newtonian program in favor of a genuine field theory of electromagnetism. In Maxwell's theory the field is at least as physically real as the electric charges or magnetic poles with which it interacts. The electromagnetic field possesses both potential and kinetic energy quite apart from the actual presence of electric charges or magnetic poles; in fact, one of the most striking manifestations of the electromagnetic field, namely, electromagnetic radiation, occurs in space completely devoid of ordinary matter with its electrical and magnetic properties. The Maxwell-Lorentz equations of the electromagnetic field—assumed to be everywhere valid *in vacuo* as well as in material media—may be written as follows (using electrostatic units):

$$\text{div } \boldsymbol{E} = 4\pi\rho \; ,$$

$$\text{curl } \boldsymbol{E} + \frac{1}{c}\frac{\partial \boldsymbol{H}}{\partial t} = 0 \; ,$$

$$\text{div } \boldsymbol{H} = 0 \; ,$$

$$\text{curl } \boldsymbol{H} - \frac{1}{c}\frac{\partial \boldsymbol{E}}{\partial t} = \frac{4\pi}{c}\,\boldsymbol{i} \; , \tag{27}$$

where $\boldsymbol{E}$, $\boldsymbol{H}$, and $\boldsymbol{i}$ are three-dimensional vector functions of x, y, z, and t, $\boldsymbol{E}$ being the electric field intensity, $\boldsymbol{H}$ the magnetic field intensity, and $\boldsymbol{i}$ the electric current density; ρ is a scalar function of x, y, z, t, the electric charge density; c is the velocity of light *in vacuo;* and div and curl have their usual vector analytical meanings. In empty space, ρ and $\boldsymbol{i}$ vanish. The laws of Ampère and Faraday are directly confirmable experimental consequences of the Maxwell-Lorentz equations. A further consequence of these equations is the existence of electromagnetic waves propagated *in vacuo* with speed c. Also, the conservation law for charge and current density can be derived from the first

---

[3] Cf. *ibid.*, p. 30: "If the three components (X, Y, Z) of the force acting on a single free particle are given functions of the coordinates (x, y, z) of the particle, they are said to define a *field of force*."

and last of equations (27) (cf. equation [7.5]). The energy of the electromagnetic field is given by the following volume integral:

$$\int_\tau \frac{E^2 + H^2}{8\pi}\, d\tau.$$

Maxwell interpreted $E^2/8\pi$ as potential energy (due to elasticity of the ethereal medium), and $H^2/8\pi$ as kinetic energy (due to motion of the ethereal medium).

In order to determine the behavior of charged particles (e.g., electrons) in electromagnetic fields one must introduce a law governing the interaction of the electromagnetic field with charged particles (whose accelerations are, indeed, the only means available for measuring the electromagnetic field). Such a law of force (sometimes called a "ponderomotive" law) was formulated by Lorentz:

$$m\dot{u} = e\left(E + \frac{u}{c} \times H\right), \tag{28}$$

where m is the mass of the particle, u is the velocity of the particle, c is the speed of light, and e is the electric charge of the particle. E and H in equation (28) must be construed as those portions of the electric and magnetic field intensities which would exist in the absence of any contribution from the particle itself; the self-field of the particle becomes infinite at the point where the charge is located, and hence the particle itself represents a singularity in the field. The necessity for omitting the self-field in computing the force on a charged particle points to an imperfection or incompleteness in the Maxwell-Lorentz field theory. A pure field theory would not treat particles as singularities (or "exceptions") but rather would somehow account for just such discrete phenomena as the existence of individual particles. Also, a pure field theory would not require a ponderomotive law over and above the field equations themselves to account for interaction of particles with the field.

The preceding discussion can be further clarified by contrasting certain mathematical features of *particle* (or *mechanical*) theories and *field* theories. Characteristic of particle theories are total differential equations with t as the only independent variable and with x, y, z, $\dot{x}$, $\dot{y}$, $\dot{z}$ (which define the trajectories of individual particles) as functions of t. Characteristic of field theories are partial differential equations with the four independent variables x, y, z, t, and with the so-called field-variables (e.g., E and H) as functions of x, y, z, t. Now, on the basis of some general properties of the solutions to partial differential equations, it is easy to see that any field theory formulated by means of *linear* partial differential equations will be unable to dispense with a ponderomotive law (expressed, for example, by a total differential equation with t as independent variable). The reason is that the linear combination (say, by addition or subtraction) of two solutions to a set of linear field equations gives rise in general to a new solution. Suppose, for example, that one has obtained two independent solutions to a set of linear field equations, each solution representing the behavior of an individual particle. Then the superposition of the two solutions is again a solution, which implies, physically, that the two particles do not interact at all (since they behave exactly the same alone as in the presence of the other particle). Non-linear field equations could, in principle at least, provide for the interaction of particles and field, and—indirectly, by means of the field—for the interaction of particles with each other without the intervention of special ponderomotive laws.

2. *Motion, Gravitation, and Electromagnetism in Special Relativity Physics*

The covariance properties of the Maxwell-Lorentz equations (27) were of crucial importance in the development of the special theory of relativity. Indeed, the fact that the Maxwell-Lorentz equations are Lorentz-covariant was one of Einstein's main reasons for believing in Lorentzian rather than Galilean kinematics. To show that the Maxwell-Lorentz equations are Lorentz-covariant one must find a set of transformation equations for E, H, i, and ρ such that the Maxwell-Lorentz equations retain their form when the space-time coordinates undergo a Lorentz transformation. Poincaré and Einstein succeeded in doing this and thereby equations (27) became a branch of special relativity. The ponderomotive law, equation (28), however, in order to be made Lorentz-covariant, must be rewritten in the form:

$$\frac{d}{dt}\left[\frac{m_0 u}{(1 - u^2/c^2)^{1/2}}\right] = e\left(E + \frac{u}{c} \times H\right), \tag{29}$$

where m_0 is now the rest mass of the particle.

On the other hand, the classical laws of motion and the classical law of gravitation are not Lorentz-covariant, so that neither can be incorporated without modification in special relativity. (The non-Lorentz-covariance of Newton's laws of motion and gravitation is evident from the assumption that mass is invariant in the Newtonian law of force, the assumption that action and reaction are absolutely simultaneous in the principle of action and reaction, the assumption that the distance in Newton's law of gravitation is invariant, and the absence of the time-coordinate in Poisson's equation.) The classical mechanics of material particles (in both the Newtonian and the analytical forms) and of continuous material media have been successfully reformulated in (special) relativistic form; the method of doing this in the case of a single particle will now be briefly explained.

On the analogy of Newton's second law of motion we may define the force acting on a particle of (relativistic) mass m as the rate of change of the momentum of the particle with time, that is,

$$f = \frac{d}{dt}(mu) = \frac{d}{dt}\left[\frac{m_0 u}{(1 - u^2/c^2)^{1/2}}\right], \tag{30}$$

where m_0 is the rest mass of the particle, the 3-vector u $(= [\dot{x}, \dot{y}, \dot{z}])$ is the velocity of the particle, and the 3-vector f is the force acting on the particle. This definition of force has the advantage of making the analogue of Newton's third law of motion, the equality of action and reaction, valid in special relativity mechanics. However, f in equation (30) is not in general parallel to acceleration, nor is f Lorentz-covariant.

In order to define a concept of "force" which *is* Lorentz-covariant, it is clear that 3-vectors must be replaced by 4-vectors. To accomplish this we first introduce the concept of "proper time." The expression for interval in special relativity, it will be recalled, is given by

$$d s^2 = - (d x^2 + d y^2 + d z^2) + c^2 dt^2 . \tag{2.3}$$

Now, along the space-time path of a moving particle, ds^2 is always positive so that ds is always real; hence we can define the *proper time* (or *world-time*) of a moving particle by means of the arc length ds along its path in space-time. For convenience, we change our space-time units and define proper time, ds, as follows:

$$d s^2 = dt^2 - \frac{1}{c^2} (d x^2 + dy^2 + d z^2) . \tag{31}$$

Proper time is identical with the time measured by a clock moving along with the particle, for in a coordinate-system at rest with respect to the particle (call it x', y', z', t'), we have

$$d s^2 = dt'^2 .$$

By equation (31), the relation between proper time ds and coordinate time dt is easily shown to be:

$$\frac{d s}{dt} = (1 - u^2 / c^2)^{1/2} , \tag{32}$$

where $u^2 = \dot{x}^2 + \dot{y}^2 + \dot{z}^2$. Both ds and s are Lorentz-invariant.

We can now define four-dimensional, Lorentz-covariant analogues of the ordinary velocity and momentum 3-vectors. The *world-velocity* U is defined by the components $U_1 = dx/ds$, $U_2 = dy/ds$, $U_3 = dz/ds$, $U_4 = dt/ds$. Using equation (32) we get:

$$\left. \begin{array}{l} U_1 = \dfrac{\dot{x}}{(1 - u^2 / c^2)^{1/2}}, \\[2mm] U_2 = \dfrac{\dot{y}}{(1 - u^2 / c^2)^{1/2}}, \\[2mm] U_3 = \dfrac{\dot{z}}{(1 - u^2 / c^2)^{1/2}}, \\[2mm] U_4 = \dfrac{1}{(1 - u^2 / c^2)^{1/2}}. \end{array} \right\} \tag{33}$$

The *world-momentum* P (also known as the *energy-momentum vector*) is defined in terms of U as follows: $P_1 = m_0 U_1$, $P_2 = m_0 U_2$, $P_3 = m_0 U_3$, $P_4 =$

$m_0 c^2 U_4$. Finally, the *world-force* (or *Minkowski force*) F is defined as the rate of change of the world-momentum with respect to proper time:

$$F = \frac{dP}{ds}.$$ (34)

In virtue of equations (33), the components of F become

$$\left. \begin{array}{l} F_1 = \dfrac{1}{(1 - u^2/c^2)^{1/2}} \dfrac{d}{dt} \dfrac{m_0 \dot{x}}{(1 - u^2/c^2)^{1/2}}, \\[2ex] F_2 = \dfrac{1}{(1 - u^2/c^2)^{1/2}} \dfrac{d}{dt} \dfrac{m_0 \dot{y}}{(1 - u^2/c^2)^{1/2}}, \\[2ex] F_3 = \dfrac{1}{(1 - u^2/c^2)^{1/2}} \dfrac{d}{dt} \dfrac{m_0 \dot{z}}{(1 - u^2/c^2)^{1/2}}, \\[2ex] F_4 = \dfrac{1}{(1 - u^2/c^2)^{1/2}} \dfrac{d}{dt} \dfrac{m_0 c^2}{(1 - u^2/c^2)^{1/2}}. \end{array} \right\}$$ (35)

The first three components of F can be formulated in terms of the components of f, as follows:

$$\left. \begin{array}{l} F_1 = \dfrac{f_x}{(1 - u^2/c^2)^{1/2}}, \\[2ex] F_2 = \dfrac{f_y}{(1 - u^2/c^2)^{1/2}}, \\[2ex] F_3 = \dfrac{f_z}{(1 - u^2/c^2)^{1/2}}, \end{array} \right\}$$ (36)

and by recalling the relation

$$E = \frac{m_0 c^2}{(1 - u^2/c^2)^{1/2}},$$ (2.6)

the last component of F can be written as:

$$F_4 = \frac{1}{(1 - u^2/c^2)^{1/2}} \frac{dE}{dt}.$$ (37)

The constancy of the first three components of F corresponds, according to equation (36), to the constancy of the relativistic force f, while the constancy of F_4 corresponds, according to equation (37), to the constancy of the relativistic total energy E. Thus, for a system of interacting particles the constancy of the vector sum of the F's for all of the particles corresponds to the relativistic conservation principles for momentum, mass, and energy. For small values of u/c, the first three components of F approach, respectively, $m_0\ddot{x}$, $m_0\ddot{y}$, and $m_0\ddot{z}$; while F_4 approaches dT/dt, i.e., the rate of increase of kinetic energy or the negative of the work performed per unit time. Thus, for small values of u/c, equations (36) and (37) reduce respectively to the Newtonian laws of motion and the classical conservation of (mechanical) energy principle.

We shall next consider briefly the problem of obtaining a Lagrangian formulation of relativistic mechanics.[4] Corresponding to the two definitions of relativistic force there are two possible approaches to our problem. One approach is simply to find an expression for L, the Lagrangian, such that the resulting Euler-Lagrange equations are identical with equations (30). The appropriate definition of L (no longer simply $T - V$) for the case of a single particle acted on by conservative forces independent of velocity turns out to be

$$L = - m_0 c^2 (1 - u^2/c^2)^{1/2} - V ,$$

where V is the potential energy depending only on position. We then have

$$\frac{\partial L}{\partial \dot{x}} = \frac{m_0 \dot{x}}{(1 - u^2/c^2)^{1/2}} ,$$

$$\frac{\partial L}{\partial \dot{y}} = \frac{m_0 \dot{y}}{(1 - u^2/c^2)^{1/2}} ,$$

$$\frac{\partial L}{\partial \dot{z}} = \frac{m_0 \dot{z}}{(1 - u^2/c^2)^{1/2}} ,$$

which gives for the Euler-Lagrange equations:

$$\frac{d}{dt} \frac{m_0 \dot{x}}{(1 - u^2/c^2)^{1/2}} = - \frac{\partial V}{\partial x} = f_x ,$$

$$\frac{d}{dt} \frac{m_0 \dot{y}}{(1 - u^2/c^2)^{1/2}} = - \frac{\partial V}{\partial y} = f_y ,$$

$$\frac{d}{dt} \frac{m_0 \dot{z}}{(1 - u^2/c^2)^{1/2}} = - \frac{\partial V}{\partial z} = f_z .$$

These equations are identical with equations (30).

The above Lagrangian formulation may be readily extended to systems of many particles, to generalized coordinates, and to velocity-dependent potential functions. However, these formulations are not four-dimensional, hence not Lorentz-covariant. It would be desirable to find a relativistic Lagrangian L' which was a world scalar, invariant with respect to Lorentz transformations. Such a Lagrangian can be found only if the forces involved (and hence the potential functions) can be expressed in covariant form. Hamilton's principle would then be written as follows:

$$\delta \int_{s_1}^{s_2} L' (x, y, z, t, U_1, U_2, U_3, U_4, s) \, d s = 0 ,$$

where U_1, U_2, U_3, U_4 are the components of the world-velocity and s is the proper time. The Euler-Lagrange equations would then look like this:

$$\frac{d}{ds} \frac{\partial L'}{\partial U_1} - \frac{\partial L'}{\partial x} = 0 ,$$

---

[4] For details see Goldstein, *Classical Mechanics*, pp. 205–11.

$$\frac{d}{ds}\frac{\partial L'}{\partial U_2} - \frac{\partial L'}{\partial y} = 0,$$

$$\frac{d}{ds}\frac{\partial L'}{\partial U_3} - \frac{\partial L'}{\partial z} = 0,$$

$$\frac{d}{ds}\frac{\partial L'}{\partial U_4} - \frac{\partial L'}{\partial t} = 0.$$

It is obvious that classical gravitational forces are not Lorentz-covariant in form and that classical electromagnetic forces automatically possess such a form. We will consider only the simplest case, that of a free particle. Suppose that, on the analogy of the classical treatment of a free particle (where $L = T$), we define L' as $\frac{1}{2}m_0U^2$; we then get

$$\frac{\partial L'}{\partial x} = \frac{\partial L'}{\partial y} = \frac{\partial L'}{\partial z} = \frac{\partial L'}{\partial t} = 0,$$

and

$$\frac{\partial L'}{\partial U_1} = m_0U_1,$$

$$\frac{\partial L'}{\partial U_2} = m_0U_2,$$

$$\frac{\partial L'}{\partial U_3} = m_0U_3,$$

$$\frac{\partial L'}{\partial U_4} = m_0U_4.$$

The Euler-Lagrange equations reduce to:

$$\frac{d}{ds}\frac{\partial L'}{\partial U_1} = \frac{d}{ds}\,m_0U_1 = 0,$$

$$\frac{d}{ds}\frac{\partial L'}{\partial U_2} = \frac{d}{ds}\,m_0U_2 = 0,$$

$$\frac{d}{ds}\frac{\partial L'}{\partial U_3} = \frac{d}{ds}\,m_0U_3 = 0,$$

$$\frac{d}{ds}\frac{\partial L'}{\partial U_4} = \frac{d}{ds}\,m_0U_4 = 0.$$

These equations are identical with equations (34) for the case of a free particle, i.e., when $F_1 = F_2 = F_3 = F_4 = 0$. (Incidentally, this result can be obtained also with other choices for the definition of L'.)

The treatment of rotational motion in classical mechanics can be carried over, with the necessary changes, into special relativity mechanics. For example, it can be shown[5] that the centrifugal force acting on a particle which is

[5] See H. Weyl, *Space-Time-Matter*, p. 224.

rotating uniformly with respect to an inertial system K is given by the expression

$$\frac{m_0 \omega \times (\omega \times r)}{1 - u^2 / c^2},$$

where m_0 is the rest mass of the particle in K, ω is the angular velocity of the particle with respect to K, r is the position vector of the particle in K, and $u = \omega \times r$.

We have seen that there are relativistic analogues of many of the main concepts of classical mechanics; the important classical concept of a rigid body, however, seems to have no relativistic analogue.[6] As for the law of gravitation, most physicists have followed Einstein in developing the special theory of relativity without reference to gravitational phenomena, which are treated instead by the general theory of relativity. Whitehead has made what appears to be one of the most successful attempts to include gravitation in special relativity.[7]

3. Einstein's General Theory of Relativity

The argument and results of Einstein's general theory of relativity may be summarized as follows.[8]

1. Einstein emphasizes the fact that his general theory of relativity is squarely based on his special theory. That the general theory is a natural and inevitable generalization and extension of the special theory may be seen, Einstein holds, from two arguments. First, epistemological considerations demand that any alleged physical cause or effect be observable. This requirement is violated by both classical and special relativity mechanics when certain preferred coordinate-systems (Galilean or inertial coordinate-systems) are held to be causes responsible for a certain class of effects, namely, the forces associated with absolute rotation. Second, the phenomenon of gravitation, and in particular the fact that all material bodies experience the same acceleration in a gravitational field (or: "inertial mass" is equivalent to "gravitational mass") suggests the possibility of assuming the equivalence of accelerated coordinate-systems and (homogeneous) gravitational fields.

Thus we arrive at the two fundamental principles of Einstein's general theory

[6] See J. L. Synge, *Relativity: The Special Theory*, pp. 35–37.

[7] Another such attempt is that by G. D. Birkhoff, "Matter, Electricity, and Gravitation in Flat Space-Time."

[8] My exposition follows closely Einstein's original papers of 1911 ("On the Influence of Gravitation on the Propagation of Light") and 1916 ("The Foundation of the General Theory of Relativity"), both available in H. A. Lorentz *et al.*, *The Principle of Relativity*.

An excellent presentation, which avoids mathematical details, may be found in E. Freundlich's monograph, *The Foundations of Einstein's Theory of Gravitation*. The Appendix, consisting of historical notes and bibliographical references, is especially valuable. Further details may be found in any of the advanced treatises mentioned in chap. ii, n. 1.

of relativity: (*a*) *the postulate of (general) relativity:* the general laws of physics must be formulated in such a way that they are valid in all coordinate-systems in arbitrary states of motion; (*b*) *the principle of equivalence:* a coordinate-system K with respect to which a homogeneous gravitational field exists (with gravitational acceleration g) is physically completely equivalent to a coordinate-system K' free of gravitational fields but uniformly accelerated with respect to K (with acceleration $-g$).

2. The postulate of general relativity is vague and, it turns out, inapplicable as stated, because coordinate-systems are no longer definable (as in the special theory) by rigid rods and standard clocks. Einstein formulates instead *the principle of covariance:* the general laws of physics are to be expressed by generally covariant equations, that is, by equations which retain their form under arbitrary transformations of the coordinates. (From the mathematical point of view the transformations are not completely arbitrary: among other conditions it is assumed that the transformation functions are continuous and differentiable.)

3. We now assume, in accordance with the principle of equivalence, that for infinitely small four-dimensional regions of space-time the special theory of relativity is always valid—provided a suitable "local" coordinate-system is chosen. (A suitable coordinate-system in this connection means one whose acceleration is such that no gravitational field appears.)

4. The interval in the local coordinate-system may be written in the form:

$$d s^2 = - d x_1^2 - d x_2^2 - d x_3^2 + d x_4^2 ,$$

where x_1, x_2, and x_3 represent the three spatial coordinates and x_4 represents the temporal coordinate, the units of space and time having been chosen so that the speed of light is unity. The coordinate-differentials of any arbitrarily selected coordinate-system can be shown to be linear homogeneous functions of dx_1, dx_2, dx_3, dx_4 for the infinitesimal four-dimensional region with which we started, from which it follows that the interval can be written in the Riemannian form (cf. above, p. 130):

$$d s^2 = \sum_{\nu=1}^{4} \sum_{\mu=1}^{4} g_{\mu\nu} d x_\mu d x_\nu , \tag{38}$$

where the $g_{\mu\nu}$ are in general functions of the x_μ, with $g_{\mu\nu} = g_{\nu\mu}$. Thus there are sixteen terms in the double summation, of which twelve are equal in pairs, hence ten independent terms.

5. Einstein now argues that the $g_{\mu\nu}$ are to be interpreted physically as the quantities which describe the gravitational field which exists relative to a particular coordinate-system. This conclusion is made plausible by the following considerations. Suppose we assume that one of the coordinate-systems of special relativity is valid throughout some finite region of space-time. As we turn from this coordinate-system with no gravitational field and constant $g_{\mu\nu}$ (in fact, $g_{11} = g_{22} = g_{33} = -1$, $g_{44} = +1$, $g_{\mu\nu} = 0$ for $\mu \neq \nu$) to any other arbitrarily

chosen coordinate-system, the $g_{\mu\nu}$ become functions of space and time, and the motion of a free particle changes from uniform to non-uniform (though the same for all particles because of the equality of gravitational and inertial mass). This latter accelerated motion we choose to interpret as due to the influence of a gravitational field; in the general case when it is impossible to introduce a special relativity coordinate-system throughout some finite region of space-time, Einstein simply *postulates* that the $g_{\mu\nu}$ describe the gravitational field. Thus gravitation is assumed to be exceptional among all known physical forces in that the $g_{\mu\nu}$ which define the gravitational field at the same time define the metrical properties of space.

6. Our next problem is to find generally covariant equations for the law of motion and for the law of gravitation (it should already be obvious that in the nature of the case these two sets of laws will be very intimately connected through the medium of the $g_{\mu\nu}$). It is to solve this problem that Einstein appeals to the theory of tensors, which may be briefly characterized as the study of sets of mathematical expressions (functions of the coordinates) which transform from one arbitrarily selected coordinate-system to another in a specified way (namely, the transformation equations for each of these functions—or "components"—are linear and homogeneous). It follows that if a proposed law of nature is expressed by equating all the components of some tensor to zero, then the law must be generally covariant (since if all components of the tensor vanish in one coordinate-system, they must vanish in all coordinate-systems). (See Appendix III for a brief account of tensors.)

7. The astonishing thing now is that if one demands that the sought-for laws of motion and gravitation satisfy certain very general and plausible conditions, then it turns out that the forms of these laws are uniquely determined. Einstein begins by noting that since the expression for the interval in generalized coordinates, given by equation (38), is invariant and since the dx_μ constitute a contravariant vector, it follows that $g_{\mu\nu}$ is a covariant tensor of rank 2; also $g_{\mu\nu}$ is symmetrical. Einstein calls $g_{\mu\nu}$ the "fundamental tensor" (it is also sometimes called the "metric tensor"). Also, a symmetrical contravariant tensor $g^{\mu\nu}$ may be defined in terms of $g_{\mu\nu}$ (see Appendix III).

8. Einstein next chooses as the law of motion for a mass-point in a gravitational field the (generally covariant) equation,

$$\delta \int_P^{P'} d\,s = 0, \tag{39}$$

which amounts to saying that a particle moving between the points (in space-time) P and P' always moves along the geodesic (or "straightest" path in space-time) joining these two points. This law is obviously satisfied by "inertial" motions in classical and special relativity physics; Einstein extends the law to motions of particles in gravitational fields. Equation (39) leads to a set of four (generally covariant) differential equations which define the geodesic. Employing—as we shall throughout the remainder of this chapter—the usual tensor summation convention (i.e., any index which is repeated in a given term is summed over all possible values of the index), we may write these differential equations as follows:

$$\frac{d^2 x_\tau}{d s^2} = - \{ \mu\nu, \ \tau \} \frac{d x_\mu}{d s} \frac{d x_\nu}{d s} \ \ (\tau = 1, 2, 3, 4) \ , \quad (40)$$

where $\{ \mu\nu, \tau \}$ is a function of the $g_{\mu\nu}$ and their first derivatives (see Appendix III). By analogy with Newtonian gravitational theory one may think of the left-hand side of equation (40) as the inertial term and of the right-hand side as the expression for gravitational force. In the absence of gravitational force (i.e., when the $g_{\mu\nu}$ and $g^{\mu\nu}$ take on the constant values characteristic of the space-time of special relativity), equations (40) reduce to:

$$\frac{d^2 x_\tau}{d s^2} = 0 \qquad (\tau = 1, 2, 3, 4) \ , \quad (41)$$

which are the equations of a straight line in space-time.

The equation for the motion of light-rays in a gravitational field follows immediately as a generalization of the equation for the motion of light-rays in empty space (i.e., flat space-time). In the special theory of relativity light-rays move according to the equation:

$$d s^2 = - d x_1^2 - d x_2^2 - d x_3^2 + d x_4^2 = 0 \ . \quad (42)$$

In the general theory of relativity, then, Einstein takes as the law of motion for light-rays:

$$d s^2 = g_{\mu\nu} d x_\mu d x_\nu = 0 \ . \quad (43)$$

9. To discover the field equations of gravitation in empty space[9] Einstein proceeds as follows. He looks for that tensor which can be formed, by differentiation, from the fundamental tensor *alone*. This tensor turns out to be a mixed tensor of fourth rank involving the $g_{\mu\nu}$ and their derivatives; it is known as the "Riemann-Christoffel tensor" and may be symbolized by $R^\rho_{\mu\nu\sigma}$. In differential geometry it is shown that the vanishing of this tensor is a necessary and sufficient condition that, by an appropriate choice of coordinate-system, the $g_{\mu\nu}$ may be constants (which means that the continuum characterized by the vanishing of this tensor is of zero curvature, i.e., the flat space-time of special relativity). Thus an initial requirement for the gravitational field equations is that they hold when $R^\rho_{\mu\nu\sigma}$ vanishes (i.e., in the absence of any gravitational field). But it is clearly too strong to require that the Riemann-Christoffel tensor *always* vanish, since it is impossible to introduce a coordinate-system with constant $g_{\mu\nu}$ throughout a finite region of a gravitational field. Hence, Einstein proceeds to introduce a new tensor (of second rank) derived by contraction from $R^\rho_{\mu\nu\sigma}$: the "contracted Riemann-Christoffel tensor," symbolized

[9] Here "empty space" means space free of matter, where "matter" includes *both* ordinary matter and also the electromagnetic field. In short, Einstein uses the term "matter" to designate everything but the gravitational field.

by $R_{\mu\nu}$. Now, if $R^\rho_{\mu\nu\sigma}$ vanishes in some continuum, so does $R_{\mu\nu}$ (this is obvious, since each component of $R_{\mu\nu}$ is simply a sum of certain components of $R^\rho_{\mu\nu\sigma}$); but not conversely. Thus, Einstein writes as his law of gravitation in empty space:

$$R_{\mu\nu} = 0 . \tag{44}$$

Since $R_{\mu\nu}$ is a symmetrical tensor, equation (44) represents a set of ten independent equations, namely, the following set of ten field equations for the ten (independent) quantities $g_{\mu\nu}$ (sometimes called the "gravitational potentials" because of the analogy between their role in the field equations and the role of ψ in Laplace's equation):

$$\{\mu\sigma, a\}\{a\nu, \sigma\} - \{\mu\nu, a\}\{a\sigma, \sigma\} + \frac{\partial}{\partial x_\nu}\{\mu\sigma, \sigma\} - \frac{\partial}{\partial x_\sigma}\{\mu\nu, \sigma\} = 0 . \tag{45}$$

Einstein next shows that purely formal considerations suffice to deduce his gravitational field equations. Assuming that the fundamental tensor $g_{\mu\nu}$ must play a central role in the law of gravitation, we demand (on the basis of analogy with the classical law of gravitation expressed in differential form, i.e., Poisson's or Laplace's equation) that the new law involve a tensor of second rank which is formed as a combination of $g_{\mu\nu}$ and its first and second derivatives and which is linear and homogeneous in the second derivatives. There is just one such tensor possible, namely, the contracted Riemann-Christoffel tensor, $R_{\mu\nu}$.

10. The solution of specific problems by means of the ponderomotive equations (40) and the field equations (45) presents formidable mathematical difficulties due to the fact that neither is linear in the $g_{\mu\nu}$ and their derivatives. (By contrast, for example, the Lorentz ponderomotive law for the motion of charged particles, equation [29], is linear in the field intensities E and H, while Laplace's equation [26] is linear in the second derivatives of the gravitational potential ψ.) In the face of these mathematical difficulties two approaches are available: one may try to obtain solutions to linear approximations of the field equations for the case of weak gravitational fields with consequently almost flat metrics (e.g., in celestial mechanics where it is known that Newton's linear theory of gravitation accounts for a large range of phenomena with great accuracy); or, one may try to obtain rigorous solutions of the field equations for special cases in which the number of independent variables is reduced by symmetry conditions.

In finding linear approximation solutions one assumes that the bodies whose motions are being investigated are so small as to make only negligible contributions to the gravitational field. If one assumes also that the velocities of the mass-points which produce the gravitational field are small compared with the velocity of light and therefore that the gravitational field itself changes only slowly with time, one can obtain solutions to the linearized field equations which have classical counterparts (e.g., the field of a mass-point). However, there are also solutions to the linearized field equations without classical counterparts (notably, the existence of "gravitational waves" propagated with the velocity of light, whose predicted intensity shows them to be too weak to be observed with available techniques).

The most important rigorous solution of the field equations known at present is that due to Schwarzschild for the case of a mass-point at rest. By imposing the conditions that the solution possess spherical symmetry, that it be static (i.e., independent of time), and that it reduce to the flat metric of special relativity at infinity, one can obtain the following values for the $g_{\mu\nu}$:[10]

$$
\left.
\begin{aligned}
g_{\rho\sigma} &= -\delta_{\rho\sigma} - \frac{2Gm}{r - 2Gm}\frac{x_\rho}{r}\frac{x_\sigma}{r} \quad (\rho, \sigma = 1, 2, 3), \\
g_{\rho 4} &= 0 \quad\quad\quad\quad\quad\quad\quad\quad (\rho = 1, 2, 3), \\
g_{44} &= 1 - \frac{2Gm}{r},
\end{aligned}
\right\} \quad (46)
$$

where $r = (x_1^2 + x_2^2 + x_3^2)^{1/2}$ is the radial coordinate from the mass-point whose mass is m, x_4 is the time-coordinate, G is the Newtonian gravitational constant, and $\delta_{\rho\sigma}$ is the Kronecker delta symbol. By introducing the values for the $g_{\mu\nu}$ characteristic of the Schwarzschild solution into the equations (40) and (43), one may find the respective paths of a small particle and of a light-ray in the gravitational field of a mass-point. The resulting prediction of a small deviation from the Keplerian orbit of a particle has been confirmed by the precession of perihelion of the planet Mercury; while the resulting prediction of a small deflection in the path of a light-ray seems to be at present not quite satisfactorily confirmed in the case of light-rays (from stars) passing near the sun. Finally, light emitted by an atom in a Schwarzschild gravitational field should be shifted in frequency toward the red, and this phenomenon seems to have been confirmed by observations of solar and stellar spectra.[11]

11. The general form of the field equations of gravitation in the presence of matter is not given by the relativity postulate alone. However, a rather natural extension of the mass-energy equivalence of special relativity leads to a tensor equation whose left-hand side is identical with that of equation (44) but whose right-hand side consists of a tensor-expression, $\kappa(T_{\mu\nu} - \frac{1}{2}g_{\mu\nu}T)$, the terms in the parentheses taking the place of the mass-density σ in Poisson's equation, while κ is closely related to G, the gravitational constant of Newton's theory. The principal justification for the introduction of the energy-tensor of matter T_σ^α (from which the tensor $T_{\mu\nu}$ and the scalar T are derived) is that the resulting equation for the gravitational field,

$$
R_{\mu\nu} = -\kappa\left(T_{\mu\nu} - \tfrac{1}{2}g_{\mu\nu}T\right), \tag{47}
$$

has as a consequence equations involving the T_σ^α which may be interpreted as expressing the conservation of momentum and energy. However, the expression which must be interpreted as the gravitational field energy is not a tensor, and hence there is no unique (generally covariant) localization of energy and momentum in a gravitational field (e.g., the energy density may vanish in some

---

[10] For an account of Schwarzschild's solution see P. Bergmann, *Introduction to the Theory of Relativity*, pp. 198–203.

[11] For a critical discussion, with references, of the empirical confirmation of Einstein's theory of gravitation, see E. T. Whittaker, *A History of the Theories of Aether and Electricity*, II, 180.

coordinate-systems but not in others). Moreover, from Einstein's point of view equation (47) represents at best a temporary makeshift until a genuine field theory which includes matter can be discovered. Such a field theory would *predict* exactly what properties matter must have rather than merely take for granted the existence of matter, in phenomenological terms, as the source of the field. Thus Einstein's judgment on equation (47) is that "it resembles a building one wing of which is constructed out of fine marble (left side of the equation) and the other of low grade wood (right side of the equation). That is, the phenomenological representation of matter is only a crude substitute for a representation which would do justice to all the known properties of matter."[12]

12. The necessity for a law of motion independent of the field equations constitutes, as already mentioned (p. 174), a departure from the strict program of field physics, and hence to this extent a blemish in the field theories both of Maxwellian electrodynamics and of Einsteinian gravitation. In the case of the latter theory, however, Einstein and some of his co-workers have been able to prove (by approximational methods similar to those used in finding and solving the linearized field equations) that the motion of mass-points is really completely determined by the field equations alone, thereby making the law of motion, equation (40), dispensable.[13] The proof, as one would expect, depends essentially on the non-linearity of Einstein's field equations.[14]

13. So far our considerations have been confined to gravitational phenomena. What now can be said of the relation between the Maxwell-Lorentz theory of the electromagnetic field and the general postulates of the general theory of relativity? It turns out that, by a simple technique of redefinition (due to Minkowski—see Appendix III), the electromagnetic field equations can be expressed in generally covariant form. Also, Einstein's law of motion, equation (40), must be modified to take account of the presence of an electromagnetic field, and the energy-components of the electromagnetic field must be related to the components of T^α_σ, which were used to express the gravitational field equations (47) in the presence of matter.[15] Einstein's judgment on the

[12] A. Einstein, "Physik und Realität," p. 335. For a discussion of this point see W. H. McCrea, "On the Objective of Einstein's Work," pp. 27–28. This article contains an excellent brief account of Einstein's most important scientific contributions. On the gravitational field energy see Bergmann, *Introduction to the Theory of Relativity*, pp. 196–97; R. C. Tolman, *Relativity, Thermodynamics, and Cosmology*, pp. 224–25; W. Pauli, *Theory of Relativity*, pp. 176–77.

[13] A. Einstein, L. Infeld, and B. Hoffmann, "The Gravitational Equations and the Problem of Motion"; A. Einstein and L. Infeld, "The Gravitational Equations and the Problem of Motion, II." For a summary discussion see Bergmann, *Introduction to the Theory of Relativity*, chap. xv.

[14] For a clear, relatively untechnical discussion of the non-linearity of Einstein's field equations, see McCrea, "On the Objective of Einstein's Work," p. 20.

[15] For details on the electromagnetic field energy see Einstein, "The Foundation of the General Theory of Relativity," in Lorentz *et al.*, *The Principle of Relativity*, pp. 155–56. The

significance of this reformulation of electromagnetic theory in a general relativistic context is worth quoting:

This inclusion of the theory of electricity in the scheme of the general theory of relativity has been considered arbitrary and unsatisfactory by many theoreticians. . . . A theory in which the gravitational field and the electromagnetic field do not enter as logically distinct structures would be much preferable.[16]

14. An obvious next step in the program of relativistic field physics would be to "unify" the gravitational and the electromagnetic fields. Einstein and others have studied this question rather fully. In particular, there have been theories which increased the number of dimensions beyond four and other theories which permitted complex (in addition to the usual real) coordinate transformations. Einstein's own reaction[17] after all these attempts had more or less failed was to return to four dimensions and real coordinate transformations, but to replace the symmetrical fundamental tensor $g_{\mu\nu}$ by a non-symmetrical tensor $g_{\mu\nu} = s_{\mu\nu} + a_{\mu\nu}$, consisting of a symmetrical and an antisymmetrical part. Using a variational principle[18] Einstein derives a set of generalized field equations. The terrific mathematical problem of finding physically meaningful solutions to these equations remains. There remains also the problem of representing the corpuscular structure of matter and electricity by means of the field equations. The ideal solution to this problem, from Einstein's point of view, would be to find singularity-free solutions to the field equations which could be interpreted as representations of corpuscles. If such solutions are found, then any laws of motion over and above the field equations themselves would be unnecessary: the field equations would completely determine the behavior of the corpuscles.[19]

---

modification of the law of motion is not discussed by Einstein in the work just cited. For a statement of the modified law, see Bergmann, *Introduction to the Theory of Relativity*, p. 193, or Tolman, *Relativity, Thermodynamics, and Cosmology*, pp. 259–60.

[16] A. Einstein, *The Meaning of Relativity*, p. 98. Any physical law may be expressed in generally covariant form. But, this mathematical fact does not reduce the great heuristic value of the principle of covariance, since it is practically impossible to accomplish the covariant reformulation for many classical physical laws (e.g., Newton's law of gravitation). On the other hand, as Tolman points out, "it would be wrong to assume, as has sometimes been done with unfortunate results in the past, that we must limit ourselves to the use of tensor equations in investigating the fundamental principles of physics" (*Relativity, Thermodynamics, and Cosmology*, p. 169; see pp. 166–74 for a general discussion of the principle of covariance).

[17] See P. A. Schilpp, *Albert Einstein: Philosopher-Scientist*, "Autobiographical Notes," pp. 89–95. Einstein's final formulation of his unified field theory occurs in *The Meaning of Relativity*, Appendix II, "Relativistic Theory of the Non-Symmetric Field."

[18] The procedure is entirely analogous to Einstein's derivation of the gravitational field equations from the variational principle:

$$\delta \int g^{\mu\nu} \Gamma^{\alpha}_{\mu\beta} \Gamma^{\beta}_{\nu\alpha} \, dx_1 \, dx_2 \, dx_3 \, dx_4 = 0 \ .$$

For details on the latter derivation see Einstein, "Hamilton's Principle and the General Theory of Relativity," in Lorentz *et al.*, *The Principle of Relativity*, pp. 145–46.

[19] See Einstein, "Physik und Realität," p. 336.

CHAPTER IX

WHITEHEAD'S
THEORY OF
RELATIVITY

1. Geometry and Tensors

Just as Einstein bases his general theory of relativity directly on his special theory of relativity, so Whitehead bases his theory of relativity[1] on the prior results of his development of geometry and kinematics by means of the method of extensive abstraction. Fundamental to that method, as we have seen, is the assumption of a certain type of uniformity in the contents of sense-perception (recall the doctrine of significance and the subsequent distinction between a duration and a strain-locus). The method of extensive abstraction eventuates in an infinite set of four-dimensional, pseudo-Euclidean space-time systems, each of which can be transformed into any one of the others by a set of formulae called the Lorentz equations. These space-time systems form the starting point of Whitehead's theory of relativity. The following passage from the Preface to R sums up Whitehead's conception of space-time.

As the result of a consideration of the character of our knowledge in general, and of our knowledge of nature in particular, . . . I deduce that our experience requires and exhibits a basis of uniformity, and that in the case of nature this basis exhibits itself as the uniformity of spatio-temporal relations. This conclusion entirely cuts away the casual heterogeneity of these relations which is the essential of Einstein's later theory. It is this uniformity which is essential to my outlook, and not the Euclidean geometry which I adopt as lending itself to the simplest exposition of the facts of nature. I should be very willing to believe that each permanent space is either uniformly elliptic or uniformly hyperbolic, if any observations are more simply explained by such a hypothesis.

It is inherent in my theory to maintain the old division between physics and

---

[1] My exposition in this chapter follows Whitehead's only detailed treatment (in R) of his theory of relativity. For the most part I have closely followed Whitehead's terminology and notation in order to facilitate the use of my discussion as an introduction to the study of Whitehead's own formulations in R.

geometry. Physics is the science of the contingent relations of nature and geometry expresses its uniform relatedness [R, pp. v–vi].[2]

Having provided himself with a uniform geometric framework, Whitehead next proceeds to formulate the laws according to which objects (of the special kind called "adjectives") appear at various loci in this framework. In other words, Whitehead turns, in his own terms, from "geometry" to "physics." The central concept in terms of which Whitehead undertakes to formulate his theory of relativity is that of "the physical field." Before explicating this concept, however, I shall attempt to explain exactly why the approach of Einstein's general theory of relativity is unacceptable to Whitehead, but why nevertheless he is able to adopt the powerful tensor analytical techniques employed by Einstein.

Since Whitehead refuses to admit any principle of relativity in his derivation of the Lorentz equations, it is obvious that he cannot adopt a generalized version of such a principle in his theory of relativity. Furthermore, Whitehead's separation of geometry and physics precludes any appeal to Einstein's principle of equivalence, which, in Whitehead's view, is essentially a way of illegitimately identifying a genuine physical phenomenon, gravitation, with purely geometric properties of space-time. It follows also that Whitehead must reject the use of arguments based on formal mathematical considerations, such as Einstein's heuristic "derivation" of the law of gravitation in empty space by means of tensor analytical manipulations of the fundamental tensor $g_{\mu\nu}$.[3] Unlike Einstein, Whitehead develops the theory of tensors in complete divorce from geometry (except, of course, where some special tensor possesses geometrical significance in Whitehead's own sense, namely, as characterizing the group of four-dimensional space-time frameworks of special relativity):

The theory of tensors is usually expounded under the guise of geometrical metaphors which entirely mask the type of application which I give to it in this work. For example, the whole idea of any 'fundamental tensor' is foreign to my purpose and impedes the comprehension of my applications [R, p. vi].

Thus, for Whitehead, Riemann's general theory of continuous manifolds with variable curvature, as well as the application of tensor analysis to the develop-

[2] Whitehead's suggestion that his theory could be developed in terms of uniformly curved space has been implemented by later physicists (see Appendix IV).

[3] It is worth quoting the following general statement on the role of mathematics in natural science: "Mere deductive logic, whether you clothe it in mathematical symbols and phraseology or whether you enlarge its scope into a more general symbolic technique, can never take the place of clear relevant initial concepts of the meaning of your symbols, and among symbols I include words. If you are dealing with nature, your meanings must directly relate to the immediate facts of observation. We have to analyse first the most general characteristics of things observed, and then the more casual contingent occurrences. There can be no true physical science which looks first to mathematics for the provision of a conceptual model. Such a procedure is to repeat the errors of the logicians of the middle-ages" (R, p. 39).

ment of that theory, are not essentially geometrical investigations at all. This is partly a matter of mere terminology, since Whitehead *defines* geometry as the study of the "uniform relatedness" of nature. Also at issue, however, is the proper procedure for discovering physical laws: Einstein's procedure is to express the relevant physical quantities as tensors and then attempt by formal mathematical considerations to discover a unique relationship between those tensors in the form of a generally covariant equation expressing an empirically confirmable law of nature; Whitehead's procedure is to express the relevant physical quantities as tensors and then attempt, by specializing the tensors for the set of coordinate-systems believed to be particularly relevant to the phenomenon in question, to discover an empirically confirmable law of nature. Einstein's procedure typically leads to a single mathematical equation, with no alternatives left open; Whitehead's procedure typically leads to a whole set of mathematical equations, among which one may be empirically most satisfactory.[4] The point I would stress is that Whitehead's narrow conception of geometry by no means forces him to reject the powerful mathematical methods of the tensor calculus; Whitehead merely insists that these methods be conceived as analytical devices for characterizing in a very general way possible mathematical relationships among physical quantities.

Whitehead's formulation of the theory of tensors differs significantly from Einstein's only at the early stages, when fundamentals are being discussed; their later elaborations of special tensors (like the Riemann-Christoffel tensor) are essentially the same—except, of course, that Whitehead never refers to any "geometrical" correlates of the tensors he introduces. Whitehead begins with the notion of a *coordinate-system* which is a set of four types of measurement that assign to each event-particle of space-time a unique ordered quadruple of real numbers, each of the four numbers determined by one of the four types of measurement. These four measurements are the *coordinates* of the event-particle. In a *pure* coordinate-system one of the four coordinates is the time of some given space-time system x, and the remaining coordinates are the spatial measurements of x. A *mixed* coordinate-system is one which is not pure (e.g., a coordinate-system in which one of the coordinates is the time of the system x and another of the coordinates is the time of a different system u).[5] By conven-

---

[4] Cf. Whitehead's reported remarks: "A further advantage of distinguishing between space-time relations as universally valid and physical relations as contingent is that a wider choice of possible laws of nature (e.g., of gravity) thereby becomes available, and while the one actual law of gravity must ultimately be selected from these by experiment, it is advantageous to choose that outlook on Nature which gives the greater freedom to experimental inquiry. Of the various laws consistent with this outlook Einstein's is one, but its theoretical aspect is different according as it is regarded from the point of view of one or the other of the two postulates [Einstein's and Whitehead's] under consideration." ("Discussion" following G. Temple, "A Generalisation of Professor Whitehead's Theory of Relativity," p. 193.)

[5] The significance of what Whitehead calls "mixed" coordinate-systems is discussed by E. Schrödinger (*Space-Time Structure*, pp. 85–86), who points out that apparently the first use of such systems was by K. Gödel (see the article "An Example of a New Type of Cosmological Solutions of Einstein's Field Equations of Gravitation," referred to in n. 16, chap. vi).

tion, the time-coordinate of a pure coordinate-system will always be represented by the fourth quantity of the ordered quadruple.

We now consider various kinds of physical quantities which may characterize event-particles. A character of event-particles which can be expressed by a single quantity, independent of coordinate-systems but a function of the coordinates of the event-particle in question, is said to be a *character of zero order*, or a *scalar* character. A character which can be expressed—relative to a selected coordinate-system—by an array of four quantities (each a function of the coordinates of the event-particle in question) such that each quantity bears a special relation to one of the four types of coordinate measurement, is said to be a *character of first order*. The four quantities are called the *components* of the character for the given coordinate-system. These definitions can easily be generalized for the case of a *character of nth order*, which will possess 4^n components. An example of a character of zero order is the gravitational potential according to some definite law of gravitation. Examples of characters of first order are: the array $(\dot{u}_1, \dot{u}_2, \dot{u}_3, 1)$, where (u_1, u_2, u_3, u_4) is a pure coordinate-system; and $(\partial\Phi/\partial u_1, \partial\Phi/\partial u_2, \partial\Phi/\partial u_3, \partial\Phi/\partial u_4)$, where Φ is the gravitational potential in the pure coordinate-system (u_1, u_2, u_3, u_4).

Suppose now we consider two coordinate-systems, u and x, and ask how the components of a given character, in u and x, respectively, are related to one another. A formula expressing a character of zero order may assign the same value to that character for every coordinate-system defined by a certain group (in the mathematical sense) of coordinate transformations; the formula is then said to possess *group invariance* with respect to that group of coordinate-systems (here the term "group" has been extended to refer to the set of coordinate-systems defined by a group of coordinate transformations).

In the case of a character of first order, there will in general be a group of coordinate-systems such that the transformations of the components of a specified character for any pair of coordinate-systems of the group are expressible by the same general rule. The case of special importance is when the components in any one coordinate-system are linear functions of the components in any other coordinate-system (of the group). If we let $(T_1^{(x)}, T_2^{(x)}, T_3^{(x)}, T_4^{(x)})$ and $(T_1^{(u)}, T_2^{(u)}, T_3^{(u)}, T_4^{(u)})$ represent the components of some first order character in the coordinate-systems x and u, respectively, the desired linear relation can be written as follows:

$$T_\mu^{(u)} = l_{\mu a} T_a^{(x)} \qquad (\mu = 1, 2, 3, 4), \quad (1)$$

where the coefficients $l_{\mu a}(\mu, a = 1, 2, 3, 4)$ are in general functions of the coordinates of the event-particles in question but independent of the particular values of the $T_\mu^{(x)}$, and where we have employed—as we shall continue to do throughout this chapter—the usual tensor summation convention. A first order character as thus described in any coordinate-system of a group is called a *group-tensor* for that group. If the general rule given by equation (1) holds for *any* pair of coordinate-systems, then the character as thus described in all coordinate-systems is called a *general tensor*, or simply a *tensor* (of first order). The components of any tensor in the coordinate-system u will be represented by the array:

$$\|T^{(u)}\|.$$

By further specifying the nature of the coefficients $l_{\mu a}$ in equation (1), one can define *covariant* and *contravariant* tensors (of first order), respectively:

$$T^{(u)}_{\mu} = T^{(x)}_{a} \frac{\partial x_a}{\partial u_\mu} \qquad (\mu = 1, 2, 3, 4) \,, \quad (2)$$

and

$$T^{\mu}_{(u)} = T^{a}_{(x)} \frac{\partial u_\mu}{\partial x_a} \qquad (\mu = 1, 2, 3, 4) \,. \quad (3)$$

So far we have considered the tensor description of physical characters. Now we introduce the notion of (group-) *tensor-invariance*, which refers to the persistence of the same mathematical formulae upon transformation from one coordinate-system to another (of a certain group of coordinate-systems) by means of the appropriate transformation equations. Thus, if in equation (2), $T^{(u)}_{\mu}$ and $T^{(x)}_{a}$ are the same formulae with the u-coordinates and the x-coordinates simply interchanged, then the formulae represented by either $T^{(u)}_{\mu}$ or $T^{(x)}_{a}$ are (group-) *tensor-covariant;* and similarly equation (3) defines (group-) *tensor-contravariance*. These definitions can be generalized for the case of nth order tensors. Whitehead concludes: "It is evident that the formulae expressing a law of nature which is not known to have any particular relation to the coordinate-systems in question should have tensor-invariance" (R, p. 147).

Clearly, the above definitions of (group-) tensors and (group-) tensor-invariance coincide with Einstein's definitions only when the groups of coordinate-systems involved in Whitehead's definitions are identical with the group of *all* coordinate-systems. Thus, one special group-tensor which is of particular importance in Whitehead's theory of relativity, but which plays only a subordinate role in Einstein's general theory of relativity, is what Whitehead calls the *Galilean tensor;* this is a second order symmetrical tensor defined as follows in the coordinate-system u:

$$\left. \begin{array}{ll} G^{(u)}_{\mu\nu} = 0 & (\mu \neq \nu; \ \mu, \nu = 1, 2, 3, 4) \,, \\[2mm] G^{(u)}_{\mu\nu} = -\omega^2_\mu & (\mu = \nu = 1, 2, 3, 4) \,, \end{array} \right\} \quad (4)$$

where

$$\left. \begin{array}{ll} \omega^2_\mu = 1 & (\mu = 1, 2, 3) \,, \\[2mm] \omega^2_4 = -c^2 \,. \end{array} \right\} \quad (5)$$

Those coordinate-systems which have the same Galilean tensor define a *Galilean group* of coordinate-systems. One such group is the set of coordinate-systems defined by the Lorentz equations; this set of orthogonal Cartesian coordinate-systems constitutes the *Cartesian group,* and corresponds to the *Cartesian*

Galilean tensor. (Earlier, we have referred to formulae which persist in form upon transformation from one member to another of the Cartesian group as "Lorentz-covariant.")

Let (x_1, x_2, x_3, x_4) and (p_1, p_2, p_3, p_4) be the coordinates of two event-particles in the coordinate-system x of the Cartesian group. Then it can be shown that $\|x_\mu - p_\mu\|$, $\|dx_\mu\|$, and $\|dp_\mu\|$ are contravariant Cartesian tensors of the first order. It follows that

$$\sum_\mu \omega_\mu^2 (x_\mu - p_\mu)^2 , \quad \sum_\mu \omega_\mu^2 dx_\mu^2 , \quad \sum_\mu \omega_\mu^2 dp_\mu^2 \tag{6}$$

are Cartesian invariants. Introducing now the following definitions,

$$\dot{x}_\mu = \frac{dx_\mu}{dx_4}, \qquad \dot{p}_\mu = \frac{dp_\mu}{dp_4}, \qquad (\mu = 1, 2, 3, 4) , \tag{7}$$

$$v_M^2 = \dot{x}_1^2 + \dot{x}_2^2 + \dot{x}_3^2 , \qquad v_m^2 = \dot{p}_1^2 + \dot{p}_2^2 + \dot{p}_3^2 , \tag{8}$$

$$\Omega_M = \left\{ 1 - \frac{v_M^2}{c^2} \right\}^{-1/2} , \qquad \Omega_m = \left\{ 1 - \frac{v_m^2}{c^2} \right\}^{-1/2} , \tag{9}$$

it follows from (6) that

$$\Omega_M^{-1} dx_4 \qquad \text{and} \qquad \Omega_m^{-1} dp_4 \tag{10}$$

are Cartesian invariants. By differentiating the first expression of (6) with respect to p_4, it can be shown that the following is also a Cartesian invariant:

$$\Omega_m \left\{ c (x_4 - p_4) - \xi_m \right\}, \tag{11}$$

where

$$\xi_m = \frac{1}{c} \sum_{\mu=1}^{3} (x_\mu - p_\mu) \dot{p}_\mu . \tag{12}$$

The preceding Cartesian invariants, as we shall see, are fundamental to Whitehead's theory of gravitation.

2. The Physical Field

As we have already seen, the term "field" is not used with a single meaning in theoretical physics. In its weakest sense "field" refers simply to any theory involving continuous distributions in space of the values of some physical magnitude, such as force or energy (in this sense Newton's theory of gravitation is a field theory); sometimes, at the other extreme, "field" refers to any theory in which all fundamental laws are expressed as partial differential equations with space and time as independent variables, and in which all corpuscular aspects of the phenomena covered by the theory may be derived from these fundamental laws (in this sense Einstein's theory of gravitation is the closest approach to a field theory known at present—although even it does not completely satisfy the criteria laid down). Whitehead's theory of relativity, and in particular his law of gravitation, seems to lie somewhere between the two extremes of Newton's theory of gravitation with its instantaneous action-at-a-distance and Einstein's theory of gravitation with no action-at-a-distance at all: Whitehead's theory of gravitation involves action-at-a-distance propagated with a certain finite velocity.[6] Whitehead's rejection of both the Newtonian and the Einsteinian approaches to natural philosophy stems from his emphasis on both the continuity and the atomicity of nature. Newton's conception of permanent atoms of matter persisting through absolute time in absolute space makes unintelligible the relations among matter, space, and time and, in particular, the continuity among the successive "states" of each bit of matter and among the successive instantaneous spaces. The conception of a completely unified physical field (Einstein's ultimate objective) makes unintelligible the very possibility of knowledge to finite, and therefore partially ignorant, minds.[7]

We have already seen in detail how the spatio-temporal continuity of nature can be derived from events and extension by the method of extensive abstraction; we have yet to investigate the origin of the atomicity of nature. The fact that atomicity is a characteristic of the physical field is what makes possible the relatively simple laws of theoretical physics. As Whitehead puts it, "Luckily the physical field is atomic, so far as concerns our approximate measurements" (R, p. 72). The fundamental principle here is that "atomicity implies two properties, one is the breakdown of relativity in that the atomic character is independent of the physical characters pervading the rest of nature, and the other is that we cannot completely exhibit this character without the whole corresponding region" (R, p. 72). Now, the atomicity of the physical field is associated essentially with scientific objects such as mass-particles, electrons, etc. Hence our problem is to find a limited region completely defined by, say, a mass-

[6] Cf. J. L. Synge, "Orbits and Rays in the Gravitational Field of a Finite Sphere According to the Theory of A. N. Whitehead," p. 308: "The theory of Whitehead seems to offer something between the two extremes of Newtonian theory on the one hand and the general theory of relativity on the other. It conforms to the requirement of Lorentz invariance (thus overcoming the major criticism against the Newtonian theory), but it does not reinstate the concept of force, with the equality of action and reaction, so that its range of applicability remains much lower than that of Newtonian mechanics. . . . It is not a field theory, in the sense commonly understood, but a theory involving action at a distance (propagated with the fundamental velocity c)."

[7] Cf. R, p. 73: "Observe that the practical atomicity of the physical and apparent characters is essential for the intelligibility of the apparent world to a finite mind with only partial perception. Without atomicity we could not isolate our problems; every statement would require a detailed expression of all the facts of nature. It has always been a reproach to those philosophers who emphasize the systematic relatedness of reality that they make truth impossible for us by requiring a knowledge of all as a condition for a knowledge of any. In the account of nature which I have just given you this objection is met in two ways: In so far as nature is systematically related, it is a system of uniform relatedness; and in the second place, intelligibility is preserved amid the contingency of appearance by the breakdown of relatedness which is involved in atomicity."

particle *m* in a definite location, this region to consist of the entire "sphere of influence" of *m*. This region will be the physical field due to *m*.

A rough conception of the physical field may be derived from a consideration of the general purpose which it serves in natural science, this purpose being to furnish a means for expressing those limitations on the "contingency of appearance" which are termed "laws of nature." More explicitly:

> They [laws of nature] are expressed by assuming that the apparent adjectives of the past indicate a certain distribution of character throughout events extending from the past into the future. It is further assumed that this hypothetical distribution of character in its turn expresses the possibilities of adjectives of appearance attachable to the future events. Thus the regulation of future adjectives of appearance by past adjectives of appearance is expressed by this intermediate distribution of character, indicated by the past and indicating the future.
>
> I call this intermediate distribution of character the 'physical field' [R, p. 71].

In order to find a mathematical formulation for the physical field, Whitehead naturally turns to the precisely defined concepts already available in his theory of events and his theory of objects. Thus the physical field is characterized with the help of the Minkowski space-time geometry of special relativity (as reinterpreted by Whitehead) and with the help of a certain type of uniform object.[8]

We begin with the notion of a four-dimensional continuum of event-particles with pseudo-Euclidean structure. Observers in different states of uniform relative motion will split up this continuum in different ways into a three-dimensional purely "spatial" continuum and a one-dimensional purely "temporal" continuum. Since there is an infinite multiplicity of possible subdivisions of the original four-dimensional continuum into "space" and "time," it follows that every event-particle lies in an infinite number of instantaneous spaces (or moments). Consider an event-particle *P*. The aggregate of event-particles lying in moments through *P* is the region co-present with *P*, which is bounded by the two three-dimensional surfaces consisting of all the null-tracks through *P*. *P*'s co-present region divides the four-dimensional continuum into two other regions, *P*'s *past* and *P*'s *future*. The three-dimensional boundary between *P*'s past and *P*'s co-present region is *P*'s *causal past;* the three-dimensional boundary between *P*'s future and *P*'s co-present region is *P*'s *causal future.* The part of *P*'s future exclusive of *P*'s causal future is *P*'s *kinematic future;* and the part of *P*'s past exclusive of *P*'s causal past is *P*'s *kinematic past.*[9] Whitehead now sug-

---

[8] This same concept of the physical field can be framed in terms of the categories of the philosophy of organism; see, e.g., PR, pp. 507–8: "The whole theory of the physical field is the interweaving of the individual peculiarities of actual occasions upon the background of systematic geometry. This systematic geometry expresses the most general 'substantial form' inherited throughout the vast cosmic society which constitutes the primary real potentiality conditioning concrescence."

[9] See the diagram in R, p. 31. In Fig. 2 the causal future of *O* is represented by *MON* and the kinematic future of *O* is represented by the region bounded by *MON* exclusive of the boundary itself. Cf. also Figure 22 in which the terminology is somewhat different, the distinction between causal and kinematic future being unnecessary in the metaphysical context.

gests that the physical field—that "intermediate distribution of character, indicated by the past and indicating the future"—be defined in terms of the causal future of P. In order actually to formulate the definition of the physical field, we must, of course, introduce some definite type of character located at P. For this purpose we refer back to certain results in the theory of objects.

We are looking for a type of object which can be situated in a more or less restricted region around P and which is practically independent of objects in the surrounding regions. Such objects Whitehead calls *adjectives*, the term suggesting the very simple way in which objects of this type ingress into nature. (Sense-objects, by contrast, require reference not only to a situation-event but also to a percipient event. However, since sense-objects "simulate adjectives for an observer who in his intellectual analysis of the circumstances forgets to mention himself" (R, p. 33), they may be appropriately called *pseudo-adjectives*.) *Uniform adjectives* (a species of uniform objects) are also referred to as *pervasive adjectives*. Finally, one can define an *adjectival particle* as a pervasive adjective qualifying a *historical route* (which is identical with a kinematic route as earlier defined—see above, p. 69). Now, as already noted, it is an empirical fact that certain well-known types of scientific objects do indeed possess the kind of independence of their environment which we are seeking. We proceed therefore to define the physical field of a mass-particle m located at P, as follows. First, we note that for a single definite space-time system, m at P means m at the point S_P at the time t_P. The causal future of P then means those points S_B at times t_B reached by some physical character due to m, starting from S_P at t_P and transmitted with the critical maximum velocity c. The *atomic physical field of m at P* is then defined as P's causal future together with P itself. P is called the *origin* of the field. Since the physical influence due to m must manifest itself in an effect on some other mass-particle, we introduce another mass-particle M at X and inquire into the effect of m on M.

Let the x-coordinates of P and X be, respectively, (p_μ) and (x_μ) ($\mu = 1, 2, 3, 4$). If P' and X' are event-particles respectively neighboring to P and X on the historical routes of m and M, then the x-coordinates of P' and X' will be, respectively:

$$(p_\mu + d p_\mu) \quad \text{and} \quad (x_\mu + d x_\mu) \quad (\mu = 1, 2, 3, 4).$$

Now, with respect to space-time systems which are members of the Cartesian group, an infinitesimal invariant, characteristic of kinematic elements like XX' and PP', has already been discovered above in (6), namely:

$$\left.\begin{aligned} dG^2 &= -\omega_\mu^2 du_\mu^2 \,, \\ &= G_{\mu\nu}^{(u)} du_\mu du_\nu \,, \end{aligned}\right\} \quad (13)$$

where $\|G_{\mu\nu}^{(u)}\|$ is the Galilean tensor defined by equations (4). For XX' and PP' we then have, respectively,

$$\left.\begin{aligned} dG_M^2 &= -\omega_\mu^2 d x_\mu^2 \,, \\ dG_m^2 &= -\omega_\mu^2 d p_\mu^2 \,, \end{aligned}\right\} \quad (14)$$

which may also be written

$$dG_M^2 = c^2\Omega_M^{-2}dx_4^2 \,,$$
$$dG_m^2 = c^2\Omega_m^{-2}dp_4^2 \,.$$

$$\left.\begin{array}{l}\end{array}\right\} \quad (15)$$

The problem now is to see how the kinematic element PP' of m's route affects any arbitrary kinematic element XX' of M's route. Whitehead solves this problem in two stages (corresponding rather closely to the two main stages of Einstein's general theory of relativity): first, he formulates a general law of motion (a law which is, indeed, based on a variational principle very similar to Einstein's); second, he formulates specific laws of gravitation and of electro-magnetism.

3. The Law of Motion

Whitehead's general law of motion involves a quantity called *impetus*. There can be various kinds of impetus; the assemblage of all the significant kinds of impetus defines the physical field. Whitehead mentions as the two principal kinds of impetus the *potential mass impetus* and the *potential electromagnetic impetus*. Writing the potential mass impetus along XX' as $\sqrt{(dJ^2)}$ and the potential electromagnetic impetus along XX' as dF, the total impetus along XX' realized by its pervasion by M is:

$$dI = M\sqrt{(dJ^2)} + c^{-1}EdF \,, \qquad (16)$$

where M is the (proper or rest) mass of M and E is the charge, in esu., of M. Between any two given event-particles, A and B, on the route of M, the realized impetus will be given by

$$\int_A^B \{M\sqrt{(dJ^2)} + c^{-1}EdF\} \,.$$

Whitehead asserts that if this total impetus is to be finite, $\sqrt{(dJ^2)}$ and dF must be homogeneous functions of the first degree in du_1, du_2, du_3, du_4, where (u_1, u_2, u_3, u_4) are any generalized coordinates of X.[10] Thus we write:

[10] The following considerations justify Whitehead's assertion. Suppose we make the plausible assumption that dF can be expressed as a series of the form

$$dF = a + a_\mu du_\mu + a_{\mu\nu}du_\mu du_\nu + a_{\mu\nu\rho}du_\mu du_\nu du_\rho + \ldots \,,$$

where the a's are functions of u_1, u_2, u_3, u_4. Taking u_1, u_2, and u_3 to be functions of the time, u_4, we have:

$$dF = \frac{dF}{du_4}\,du_4 = \left\{\frac{a}{du_4} + a_\mu\dot{u}_\mu + a_{\mu\nu}\dot{u}_\mu\dot{u}_\nu du_4 + a_{\mu\nu\rho}\dot{u}_\mu\dot{u}_\nu\dot{u}_\rho du_4^2 + \ldots\right\}du_4 \,.$$

Now in the integral

$$\int_{u_4 \text{ at } A}^{u_4 \text{ at } B} c^{-1}EdF \,,$$

the a term will approach infinity as du_4 approaches 0, while the $a_{\mu\nu}$, $a_{\mu\nu\rho}$, and all higher terms will approach zero as du_4 approaches zero. Hence, by ignoring all terms except $a_\mu\dot{u}_\mu$, we may

$$dJ^2 = J^{(u)}_{\mu\nu} du_\mu du_\nu ,$$
$$dF = F^{(u)}_\mu du_\mu .$$
$$\left.\begin{array}{c}\\\\\end{array}\right\} \quad (17)$$

Thus, $\|J^{(u)}_{\mu\nu}\|$ is a covariant tensor of second order which may be assumed to be symmetric, and $\|F^{(u)}_\mu\|$ is a covariant tensor of first order. Also, if we assume that (u_1, u_2, u_3, u_4) constitutes a pure coordinate-system, we may write:

$$\dot{u}_\mu = \frac{du_\mu}{du_4} \qquad (\mu = 1, 2, 3, 4) ,$$

so that dI/du_4 becomes a function of $\dot{u}_1, \dot{u}_2, \dot{u}_3$, and of u_1, u_2, u_3, u_4. Whitehead's general law of motion is obtained by integrating dI along the route of M between any two event-particles A and B, and then setting the variation of the resulting integrand equal to zero:

$$\delta \int_A^B dI = \delta \int_A^B \{ M \sqrt{(dJ^2)} + c^{-1} E dF \} = 0 . \qquad (18)$$

The result of performing the variation is a set of differential equations of the Euler-Lagrange type:[11]

$$\frac{d}{du_4} \frac{\partial}{\partial \dot{u}_\mu} \frac{dI}{du_4} - \frac{\partial}{\partial u_\mu} \frac{dI}{du_4} = 0 \qquad (\mu = 1, 2, 3) . \qquad (19)$$

Whitehead's variational principle, equation (18), is similar to Einstein's variational principle, equation (8.39), with the important difference that, whereas Einstein interprets ds as a geometrical quantity (space-time interval), Whitehead interprets dI as a physical quantity (impetus).[12] From the meta-

---

be certain of the finiteness of the integral. In other words, we conclude that dF is a linear homogeneous function of du_1, du_2, du_3, du_4. Similar considerations lead to the conclusion that dJ^2 must be homogeneous and of second degree in du_1, du_2, du_3, du_4.

[11] It is easy to see that equations (18) and (19) are only slightly modified versions of the more usual formulations of, respectively, Hamilton's principle and the Euler-Lagrange equations (see equations [8.16] and [8.17]). Thus, if we write

$$dI = L du_4 ,$$

where L is the Lagrangian function, then equation (18) becomes

$$\delta \int_A^B L du_4 = 0 ,$$

and equations (19) become

$$\frac{d}{du_4} \frac{\partial L}{\partial \dot{u}_\mu} - \frac{\partial L}{\partial u_\mu} = 0 \qquad (\mu = 1, 2, 3) .$$

Of course, as we have seen in our discussion of relativistic Lagrangian mechanics (p. 178), L is no longer equal to the difference of the kinetic and potential energies.

[12] Cf. PR, pp. 506–7: "It is usual to term an 'infinitesimal' element of this integral [in the variational principle expressing the general law of motion] by the name of an element of distance. But this name, though satisfactory as a technical phraseology, is entirely misleading. . . . it would be better—so far as explanation is concerned—to abandon the term 'distance' for this integral, and to call it by some such name as 'impetus,' suggestive of its physical import."

physical standpoint of Whitehead's philosophy of organism, the constant occurrence of variational principles in natural science may be traced to a general ontological "principle of determination," which requires that an individual (be it a man, an animal, or an electron) will tend under certain circumstances unconsciously to envisage "that one immediate possibility of attainment which represents the closest analogy to its own immediate past, having regard to the actual aspects which are there for prehension." From which it follows that "The laws of physics represent the harmonised adjustment of development which results from this unique principle of determination. Thus dynamics is dominated by a principle of least action, whose detailed character has to be learnt from observation" (SMW, p. 155). The precise physical significance of dI and therefore of the tensors $\|J_{\mu\nu}^{(u)}\|$ and $\|F_{\mu}^{(u)}\|$ in equations (17) must, then, be determined empirically. To this problem we now turn.

Equations (19) may be cast into a more perspicuous form by expressing the potential mass impetus as follows:

$$dJ^2 = dG_M^2 - g_{\mu\nu}^{(u)} du_\mu du_\nu \,, \tag{20}$$

or:

$$\left.\begin{aligned} dJ^2 &= - \omega_\mu^2 du_\mu^2 - g_{\mu\nu}^{(u)} du_\mu du_\nu \,, \\ &= G_{\mu\nu}^{(u)} du_\mu du_\nu - g_{\mu\nu}^{(u)} du_\mu du_\nu \,, \end{aligned}\right\} \tag{21}$$

where (u_1, u_2, u_3, u_4) are the generalized coordinates of M. In effect, the symmetric covariant tensor $\|J_{\mu\nu}^{(u)}\|$ has been split up into the difference of two symmetric covariant tensors; and inspection of equation (21) suggests that one of these tensors, $\|G_{\mu\nu}^{(u)}\|$, is connected with the inertial aspects of M's motion, and the other, $\|g_{\mu\nu}^{(u)}\|$, with the gravitational aspects of M's motion. Using equation (20) we can now write for the equations of motion:

$$\frac{d}{du_4}\left\{ \frac{M}{\Gamma_{(u)}} \frac{\partial}{\partial \dot{u}_\mu} \left(-\tfrac{1}{2} c^2 \Gamma_{(u)}^2 \right) \right\} - \frac{M}{\Gamma_{(u)}} \frac{\partial}{\partial u_\mu} \left(-\tfrac{1}{2} c^2 \Gamma_{(u)}^2 \right)$$
$$= E F_{\mu\rho}^{(u)} \dot{u}_\rho \qquad (\mu = 1, 2, 3)\,, \tag{22}$$

where (u_1, u_2, u_3, u_4) are any generalized coordinates of the situation of M, and

$$\Gamma_{(u)} = c^{-1} \sqrt{\frac{dJ^2}{du_4^2}},$$

$$F_{\mu\rho}^{(u)} = \frac{\partial F^{(u)}}{\partial u_\rho} - \frac{\partial F_\rho^{(u)}}{\partial u_\mu}.$$

For Cartesian coordinates (x_1, x_2, x_3, x_4) these equations become:

$$\frac{d}{dx_4} \frac{M\dot{x}_\mu}{\Gamma} + g_{\mu\rho} \frac{d}{dx_4} \frac{M\dot{x}_\rho}{\Gamma} + \frac{M}{\Gamma} g\,[\rho\,\sigma,\,\mu]\,^{(x)}\dot{x}_\rho\dot{x}_\sigma$$
$$= E F_{\mu\rho}^{(x)} \dot{x}_\rho \qquad (\mu = 1, 2, 3)\,, \tag{23}$$

where

$$g\,[\rho\sigma,\,\mu]^{(x)} = \frac{1}{2}\left(\frac{\partial \overset{(x)}{g_{\mu\rho}}}{\partial x_\sigma} + \frac{\partial \overset{(x)}{g_{\mu\sigma}}}{\partial x_\rho} - \frac{\partial \overset{(x)}{g_{\rho\sigma}}}{\partial x_\mu}\right), \tag{24}$$

and $\Gamma_{(x)}$ has been abbreviated to Γ.

It is now easy to confirm our earlier idea as to the significance of the tensors $\|G\{u\}_{\mu\nu}^{(u)}\|$ and $\|g_{\mu\nu}^{(u)}\|$. In the absence of both gravitational and electromagnetic fields, both the $F_\mu^{(x)}$ and the $g_{\mu\nu}^{(x)}$ should vanish, and equations (23) then reduce to:

$$\frac{d}{dx_4}\left(\frac{M\dot{x}_\mu}{(1 - v_M^2/c^2)^{1/2}}\right) = 0 \qquad (\mu = 1, 2, 3), \tag{25}$$

which is the special relativity law of inertia (cf. equation [8.30]). In the absence of gravitational fields, equations (23) reduce to:

$$\frac{d}{dx_4}\left[M\dot{x}_\mu\left(1 - \frac{v_M^2}{c^2}\right)^{-1/2}\right] = EF_{\mu\rho}^{(x)}\dot{x}_\rho \qquad (\mu = 1, 2, 3), \tag{26}$$

which is identical with the electromagnetic ponderomotive law of special relativity (equations [8.29]), provided we identify $(F_{14}^{(x)}, F_{24}^{(x)}, F_{34}^{(x)})$ with the electric force (in esu.) and $(cF_{23}^{(x)}, cF_{31}^{(x)}, cF_{12}^{(x)})$ with the magnetic force. Furthermore, Whitehead's definition of $\|F_{\mu\rho}^{(u)}\|$ enables him to use precisely Einstein's technique for formulating the Maxwell-Lorentz equations in covariant form (see Appendix III, pp. 234–35). The next task of Whitehead's theory of relativity is to discover a suitable law of gravitation, i.e., a suitable form for the tensor $\|g_{\mu\nu}^{(u)}\|$.

4. The Laws of Gravitation and Electromagnetism

To determine dJ^2 as a function of other mass-particles in other routes and thereby to find the expression for the gravitational field, Whitehead introduces the following considerations. The physical field of m (at P) has been defined as P's causal future together with P itself. Now, we already know that for all coordinate-systems in the Cartesian group the interval is invariant. But for any two event-particles which are causally related, it is easy to see that the interval vanishes (in other words, the two event-particles must lie on a null-track). Thus we have for the two event-particles X and P, with coordinates in the x-system, respectively, (x_1, x_2, x_3, x_4) and (p_1, p_2, p_3, p_4):

$$c^2(x_4 - p_4)^2 - [(x_1 - p_1)^2 + (x_2 - p_2)^2 + (x_3 - p_3)^2] = 0 \ ;$$

or:

$$c(x_4 - p_4) = r_{(x)}, \tag{27}$$

where $r_{(x)}$ is the spatial separation of X and P in the x-system, i.e., $r_{(x)} = [(x_1 - p_1)^2 + (x_2 - p_2)^2 + (x_3 - p_3)^2]^{1/2}$. We must somehow make use of equation (27) in expressing the law of gravitation.

Assume that

$$dJ^2 = dG_M^2 - \frac{2}{c^2}\psi_m \, dG_m^2 , \qquad (28)$$

where ψ_m represents the classical gravitational potential function for a particle m (as formulated in equation [8.22]). By making appropriate approximations, one can derive Newton's law for the gravitational force exerted on M by m. Using equations (15), equation (28) gives:

$$dJ^2 = c^2 \Omega_M^{-2} dx_4^2 - 2\psi_m \Omega_m^{-2} dp_4^2 .$$

Now, according to classical physics the gravitational force exerted on M by m is supposed to act instantaneously over the intervening space; hence $x_4 = p_4$ and $dx_4 = dp_4$, and we have:

$$\sqrt{\left|\frac{dJ^2}{dx_4^2}\right|} = \sqrt{\left[c^2 - v_M^2 - 2\psi_m \left(1 - \frac{v_m^2}{c^2}\right) \right]} .$$

Since electromagnetic fields are supposed to be absent, $M\sqrt{(dJ^2/dx_4^2)} = dI/dx_4$, and equations (19) become:

$$\frac{d}{dx_4} \frac{\partial}{\partial \dot{x}_\mu} \left(M\sqrt{\left|\frac{dJ^2}{dx_4^2}\right|} \right) - \frac{\partial}{\partial x_\mu} \left(M\sqrt{\left|\frac{dJ^2}{dx_4^2}\right|} \right) = 0 \quad (\mu = 1, 2, 3)$$

or, remembering that $v_m \ll c$ (which means that the motion of the particle producing the gravitational field is small compared with the speed of light),

$$- M\ddot{x}_\mu - M \frac{\partial \psi_m}{\partial x_\mu} = 0 \qquad (\mu = 1, 2, 3) . \quad (29)$$

These three equations represent the classical motion of M under the influence of the gravitational force acting on M due to m.

Whitehead now seeks a form for ψ_m which is analogous to the Newtonian gravitational potential function but which replaces the distance r by a Cartesian invariant (or Lorentz-invariant) analogous to r. Such an expression has already been discovered, namely,

$$\Omega_m \{ c (x_4 - p_4) - \xi_m \} . \qquad (11)$$

Taking account of the causal condition expressed by equation (27), the invariant reduces to:

$$\Omega_m \{ r_{(x)} - \xi_m \} . \qquad (30)$$

Hence Whitehead assumes

$$\psi_m = \frac{\gamma m}{\Omega_m \{ r_{(x)} - \xi_m \}} , \qquad (31)$$

where γ is a constant. Whitehead's actual law of gravitation then becomes:

$$dJ^2 = dG_M^2 - \frac{2}{c^2} \sum_m \psi_m \, dG_m^2 , \qquad (32)$$

where the summation is over all mass-particles such as m in kinematic elements such as PP', causally related to XX' (the kinematic element of M). The factor $2/c^2$ is required so that the constant γ will be identical with the Newtonian gravitational constant G in the limiting case when $v_m \ll c$.[13] Also, it will be noted that the summation Σ_m omits the self-field of the test-particle M whose motion is being considered.

The sense in which $\Omega_m\{r_{(x)} - \xi_m\}$ is analogous to the pre-relativistic distance (r) may be seen by writing it out in its unabbreviated form; this gives:

$$\frac{1}{\sqrt{(1 - v_m^2/c^2)}} \{ c\,(x_4 - p_4)\,\dot{p}_4 - \frac{1}{c}[\,(x_1 - p_1)\,\dot{p}_1 + (x_2 - p_2)\,\dot{p}_2 + (x_3 - p_3)\,\dot{p}_3]\,\}.$$

Now, it is not difficult to show that the expression just obtained can be written as the scalar product of two vectors, namely, the unit tangent vector to m's

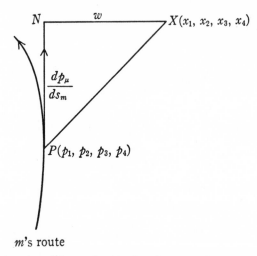

m's route

Fig. 25.—The invariant "distance" w between the route of a mass-particle m and some event-particle X in the causal future of m.

route at P (which may be written dp_μ/ds_m^2, where ds_m is defined as $dp_\mu dp_\mu$) and the vector-difference between the vectors x_μ and p_μ. The geometrical significance of this scalar product can be indicated by a diagram (see Figure 25). The situation depicted is as follows: P is an event-particle on m's route; X is an event-particle not on m's route; from X a perpendicular is dropped to the tangent at P. We then have:

$$\vec{PX} \cdot \frac{dp_\mu}{ds_m} = |\vec{PN}|,$$

and

$$(PX)^2 = (PN)^2 + (NX)^2.$$

[13] See J. L. Synge, *The Relativity Theory of A. N. Whitehead*, p. 18. Note that Synge's gravitational constant k is not identical with Whitehead's γ.

Since PX represents a portion of a null-track,

$$(PX)^2 = (x_1 - p_1)^2 + (x_2 - p_2)^2 + (x_3 - p_3)^2 - c^2 (x_4 - p_4)^2 = 0 \; ;$$

it follows that $\overrightarrow{PN}$ and $\overrightarrow{NX}$ have the same magnitude (or absolute value). Hence, with $\overrightarrow{PX} = (x_\mu - p_\mu)$,

$$| \overrightarrow{NX} | = w = \sum_\mu - \omega_\mu^2 (x_\mu - p_\mu) \frac{d p_\mu}{d s_m},$$

where the factor $-\omega_\mu^2$ is necessary to take account of the pseudo-Euclidean character of space-time (cf. the space-time metric dG^2 defined by equation [13]). Also, we note that $dp_\mu/ds_m = \dot{p}_\mu c^{-1}(1 - v_m^2/c^2)^{-1/2}$. Hence:

$$w = c^{-1}\left(1 - \frac{v_m^2}{c^2}\right)^{-1/2} \{ - (x_1 - p_1)\,\dot{p}_1 - (x_2 - p_2)\,\dot{p}_2 - (x_3 - p_3)\,\dot{p}_3$$
$$+ c^2 (x_4 - p_4)\,\dot{p}_4 \} \; ,$$

or:

$$w = \Omega_m \{ r_{(x)} - \xi_m \}. \tag{33}$$

Thus, Whitehead's formula for the gravitational potential function can be written:[14]

$$\psi_m = \frac{\gamma m}{w}. \tag{34}$$

When m is at rest, i.e., when m's route is parallel to the time-axis (x_4) in the x-system, w becomes simply the spatial distance of X from m's route. In general, the invariant w plays the role of "distance" from X to the route of m.[15]

After deriving his law of gravitation as outlined above, Whitehead points out an alternative method of derivation. One might start from the differential equation

$$\frac{\partial^2 \psi}{\partial x_1^2} + \frac{\partial^2 \psi}{\partial x_2^2} + \frac{\partial^2 \psi}{\partial x_3^2} - \frac{1}{c^2} \frac{\partial^2 \psi}{\partial x_4^2} = 0$$

[14] The tensor $\|J_{\mu\nu}^{(x)}\|$ takes on an especially clear and simple form when expressed in terms of w. Equations (28), (13), and (34) give:

$$dJ^2 = J_{\mu\nu}^{(x)} d x_\mu d x_\nu = G_{\mu\nu}^{(x)} d x_\mu d x_\nu - \frac{2}{c^2} \frac{\gamma m}{w} G_{\mu\nu}^{(x)} d p_\mu d p_\nu.$$

But

$$G_{\mu\nu}^{(x)} d p_\mu d p_\nu = \frac{1}{w^2} (x_\mu - p_\mu)(x_\nu - p_\nu)\, d x_\mu d x_\nu.$$

Hence:

$$\| J_{\mu\nu}^{(x)} \| = \| G_{\mu\nu}^{(x)} \| - \frac{2}{c^2} \frac{\gamma m}{w^3} \| x_\mu - p_\mu \| \, \| x_\nu - p_\nu \|.$$

Whitehead himself does not employ this form for $\|J_{\mu\nu}^{(x)}\|$; see, however, Synge, *The Relativity Theory of A. N. Whitehead*, p. 9.

[15] The invariant represented by (11)—which reduces to w for two event-particles on a null-track—was introduced by M. Born in his definition of rigid motions in special relativity. For further details and references, see J. L. Synge, *Relativity: The Special Theory*, pp. 29–30, 36–37.

as the only invariant form of linear differential equation of second order (and simply the four-dimensional generalization of Laplace's equation [8.26]); then, one could deduce the expression

$$\psi = \sum_m \psi_m = \sum_m \frac{\gamma\, m}{\Omega_m \{\, r_{(x)} - \xi_m \}}\,,$$

as the only invariant solution for a single point-wise discontinuity. But, Whitehead concludes, "The procedure of thought which I have adopted seems to me to be better suited to throw into relief the fundamental ideas concerning nature" (R, p. 82). The alternative to his own preferred method suggested by Whitehead is, of course, similar to the method actually employed by Einstein in deriving his law of gravitation (see above, pp. 183–84).

If we restrict ourselves to gravitational fields, then in any given problem our task is to find the proper expressions for the $g_{\mu\nu}^{(u)}$ of equation (21). For Cartesian coordinates we find from equations (20), (15), and (27):

$$g_{\mu\nu}^{(x)} d x_\mu d x_\nu = \sum_m \frac{2\gamma\, m\, (d x_4 - d\, r_{(x)}/c)^{\,2}}{\Omega_m^3\, (r_{(x)} - \xi_m)}.$$

Since $dr_{(x)} = (\partial r_{(x)}/\partial x_\mu) dx_\mu$, we get:

$$g_{\mu\nu}^{(x)} d x_\mu d x_\nu = \sum_m \frac{2\gamma\, m}{\Omega_m^3\, (r_{(x)} - \xi_m)} \bigg(d x_4^2 - \frac{1}{c}\frac{\partial\, r_{(x)}}{\partial x_\mu}\, d x_\mu d x_4$$
$$+ \frac{1}{c^2}\frac{\partial\, r_{(x)}}{\partial x_\mu}\frac{\partial\, r_{(x)}}{\partial x_\nu}\, d x_\mu d x_\nu \bigg).$$

Equating the coefficients of the corresponding terms in $dx_\mu dx_\nu$, one obtains the following values for the $g_{\mu\nu}^{(x)}$:

$$\left.\begin{array}{ll}
g_{\mu\nu}^{(x)} = \displaystyle\sum_m \frac{2\gamma\, m}{c^2\Omega_m^3\, (r_{(x)} - \xi_m)}\frac{\partial\, r_{(x)}}{\partial x_\mu}\frac{\partial\, r_{(x)}}{\partial x_\nu} & (\mu,\, \nu \neq 4)\,, \\[2ex]
g_{\mu4}^{(x)} = \displaystyle\sum_m \frac{2\gamma\, m}{c^2\Omega_m^3\, (r_{(x)} - \xi_m)}\frac{\partial\, r_{(x)}}{\partial x_\mu}\Big(\frac{\partial\, r_{(x)}}{\partial x_4} - c\Big) & (\mu \neq 4)\,, \\[2ex]
g_{44}^{(x)} = \displaystyle\sum_m \frac{2\gamma\, m}{c^2\Omega_m^3\, (r_{(x)} - \xi_m)}\Big(\frac{\partial\, r_{(x)}}{\partial x_4} - c\Big)^{\!2}. &
\end{array}\right\} \quad (35)$$

In the absence of gravitational fields, Whitehead adopts the standard Maxwell-Lorentz field equations (equations [8.27]) for electromagnetic phenomena. However, in the presence of a gravitational field, Whitehead modifies the Maxwell-Lorentz equations to take account of the influence of gravitation on electromagnetic phenomena. Two of the four equations (the second and third of equations [8.27]) remain unchanged and may be expressed in terms of the components of the tensor $\|F_{\mu\rho}\|$ (see equations [18] and [19], Appendix III). In order to formulate the other two equations (the first and fourth of equations [8.27]), Whitehead introduces a skewsymmetrical tensor, $\|(JF)_{(x)}^{\mu\nu}\|$ defined as follows:

$$\| (JF)_{(x)}^{\,\mu\nu} \| = J_{(x)}^{\rho\nu} J_{(x)}^{\sigma\nu} F_{\rho\sigma}^{(x)}\,, \tag{36}$$

where $\|J^{\mu\nu}_{(x)}\|$ is the so-called *associate* tensor of $\|J^{(x)}_{\mu\nu}\|$, a contravariant tensor defined by:

$$J^{\mu\rho}_{(x)}\,J^{(x)}_{\nu\rho}\begin{cases} = 0 & (\mu \neq \nu)\,, \\ = 1 & (\mu = \nu)\,. \end{cases}$$

Letting $\rho_{(x)}$ be the electric charge density in x-space, we find that

$$\|\,\rho_{(x)}\dot{x}_\mu\|$$

is a tensor, the so-called "contravariant electric motion tensor." Then the remaining Maxwell-Lorentz equations can be written:

$$\frac{\partial}{\partial x_a}\,(JF)^{\mu a}_{(x)} = \frac{4\pi\rho_{(x)}}{c^2}\,\dot{x}_\mu \quad (\mu = 1, 2, 3, 4)\,. \quad (37)$$

(Compare these equations with equations [21] and [22], Appendix III.) If the gravitational field is constant throughout the region of the electromagnetic field (i.e., if the $J^{(x)}_{\mu\nu}$ are constants), then equations (37) take the form:

$$J^{\sigma a}_{(x)}\,\frac{\partial F^{(x)}_{\lambda\sigma}}{\partial x_a} = \frac{4\pi\rho_{(x)}}{c^2}\,J^{(x)}_{\lambda\mu}\dot{x}_\mu \quad (\lambda = 1, 2, 3, 4)\,. \quad (38)$$

5. The Empirical Consequences of Whitehead's Theory of Relativity

Like Einstein's, Whitehead's law of gravitation reduces, in first approximation, to Newton's law of gravitation. This we have already seen, namely, the substitution in dJ^2 of the expression $(\gamma m)/r$ for the expression $(\gamma m)/\Omega_m\{r_{(x)} - \xi_m\}$ leads to Newton's law of gravitation (see equations [29]). The application of Whitehead's law of gravitation to planetary motions leads to Einstein's result on the advance of perihelion of the elliptical orbit of a mass-particle moving in a static, spherically symmetrical gravitational field (R, p. 105). Furthermore, Eddington was able to show (see Appendix IV) that for this type of gravitational field Whitehead's law of gravitation leads precisely to the Schwarzschild solution of Einstein's field equations (equations [8.46]).

We must consider next the behavior of light. In the absence of gravitational fields, of course, the Maxwell-Lorentz equations lead to the result that light is propagated in straight lines with speed c. In the presence of a gravitational field, one must resort to equations (37). If the wave-lengths of the light-waves are small compared with the linear dimensions of the regions within which the gravitational field is constant, one may use equations (38), which for uncharged space become:

$$J^{\sigma a}_{(x)}\,\frac{\partial F^{(x)}_{\mu\sigma}}{\partial x_a} = 0\,. \quad (39)$$

From these equations Whitehead derives (R, pp. 108–9) for the paths of rays of light of short wave-length the condition:

$$dJ^2 = 0 \ , \tag{40}$$

which is equivalent to Einstein's *assumption*,

$$d s^2 = g_{\mu\nu} d x_\mu d x_\nu = 0 \ .$$

As one would expect from the identity of Whitehead's expression for a static, spherically symmetrical gravitational field with the Schwarzschild solution of Einstein's field equations, Whitehead deduces from equations (39) that light passing near a massive object like the sun will be deviated in precisely the degree predicted by Einstein. Also, Whitehead shows that according to his theory of the interaction of gravitation and light any experiment of the Michelson-Morley type will give a null result (R, p. 111).

In the case of the third of the so-called "crucial" tests of the general theory of relativity—the red-shift in light emitted by atoms in a gravitational field— Whitehead's theory yields a formula for the red-shift differing from Einstein's formula by a factor of 13/6, i.e., Whitehead's theory predicts about twice as great an effect as Einstein's theory. Since Einstein's prediction seems to have been confirmed by astronomical observations, one might be tempted to conclude that Whitehead's theory had been severely discredited. It must be pointed out, however, that Whitehead deduces the red-shift in question with the aid of a simple, highly schematized model of the molecule; and that J. L. Synge, using a somewhat different approach (in fact, Einstein's approach), has succeeded in deducing from Whitehead's theory precisely Einstein's value for the red-shift.[16] Moreover, Whitehead's approach is presumably outmoded anyway, since quantum mechanical considerations can hardly be omitted in analyzing the interactions of matter and radiation. For similar reasons several other of Whitehead's calculations are probably of little interest today. He deduces, for example, two more effects of a gravitational field on spectral lines, namely, a limb effect (i.e., a shift in spectral lines toward the red as one proceeds from the center of a massive body like the sun to its rim), and a doubling or tripling of spectral lines in a strong gravitational field. Finally, Whitehead deduces (R, chap. xvi) slight modifications, due to the presence of a gravitational field, of both steady electric and steady magnetic fields. Thus, on the basis of equations (38) Whitehead shows (R, pp. 127–29) that the interaction of a steady electric field and a permanent gravitational field produces a magnetic field. The effect

[16] In R (see p. 116) Whitehead's value for the red-shift in a gravitational field differs from Einstein's value by the negligible factor 7/6. However, in his personal copy of R (which may now be consulted in the Princeton University Library) Whitehead corrected his calculation and arrived at the factor 13/6, which is cited in the text above. Whitehead also corrected his calculations for the limb effect and the doubling effect. Unfortunately, Whitehead's handwritten marginal corrections are not accompanied by any comments on their significance.

Synge's derivation of the red-shift from Whitehead's theory may be found in his *The Relativity Theory of A. N. Whitehead*, p. 45.

is an extremely minute one; the maximum magnetic field produced at the surface of the earth, for example, would be $1.2 \times 10^{-9} \times F$ gausses, where F is the electric field intensity in esu. Whitehead concludes that "an experiment of sufficient accuracy to detect the magnetic force, if it exists, would be of great interest as forming a crucial experiment to test the formula for dJ^2 here adopted" (R, p. 129). Of considerably more interest insofar as an experimental comparison of Einstein's and Whitehead's theories of gravitation is concerned is Whitehead's deduction (R, chap. xvii) of some slight corrections in the motion of the moon moving in the combined gravitational fields of earth and sun. Astronomical observations may thus be expected eventually to settle the question of which theory, Einstein's or Whitehead's, provides a more satisfactory basis for celestial motions.[17]

So far I have been expounding Whitehead's theory of relativity as if it were simply an alternative to Einstein's general theory of relativity. Actually, this is not the way in which Whitehead himself views the relation between his own theory and Einstein's. Whitehead is willing to accept Einstein's mathematical formulae for the laws of motion and gravitation (provided, of course, these formulae are reinterpreted in the terms of Whitehead's own natural philosophy):

If the above formula gives results which are discrepant with observation, it would be quite possible with my general theory of nature to adopt Einstein's formula, based upon his differential equations, for the determination of the gravitational field. They have however, as initial assumptions, the disadvantage of being difficult to solve and not linear. But it is purely a matter for experiment to decide which formula gives the small corrections which are observed in nature [R, p. 84].

It is not perfectly plain just how one should interpret Einstein's general theory of relativity in such a way as to be consonant with Whitehead's natural philosophy. Unfortunately, Whitehead does not elaborate on this point and we are compelled to guess at his meaning. A possible interpretation of Whitehead's meaning is suggested by a remark of A. S. Eddington concerning Whitehead's theory of relativity.[18] Eddington claims that Whitehead is correct in demanding a "basis of uniformity" in the formulation of physical laws but that he overlooks the fact that such uniformity is present in Einstein's general theory of relativity. This uniformity Eddington finds in the circumstance that what he calls "the ten principal coefficients of curvature" are always zero in empty space (while the other ten "coefficients of curvature" vary with the gravitational field and do not generally vanish). Eddington is referring here, in a somewhat imprecise fashion, to the fact that, according to Einstein's gravitational field equations, ten independent functions of the components of $R^\rho_{\mu\nu\sigma}$ vanish (i.e., the ten independent components of $R_{\mu\nu}$), while ten more independent functions of the $R^\rho_{\mu\nu\sigma}$

[17] Recently, A. Schild has discovered another phenomenon about which Einstein's and Whitehead's theories make conflicting predictions, namely, the motion of double stars (see Appendix IV).

[18] A. S. Eddington, *The Nature of the Physical World*, pp. 145–46.

vary arbitrarily.[19] But the purely formal mathematical uniformity to which Eddington refers is surely *not* what Whitehead has in mind. Rather, I believe, Whitehead's reinterpretation of Einstein's general theory of relativity must involve the splitting up of Einstein's fundamental tensor, $g_{\mu\nu}$, into two parts, the one consisting of a tensor with the constant components,

$$
\begin{matrix}
-1 & 0 & 0 & 0 \\
0 & -1 & 0 & 0 \\
0 & 0 & -1 & 0 \\
0 & 0 & 0 & 1 \, ,
\end{matrix}
$$

representing purely spatio-temporal or inertial properties; and the other consisting of a tensor with variable components representing the gravitational field.[20]

Should both Whitehead's and Einstein's laws of gravitation turn out to be empirically unsatisfactory, it would become necessary to investigate other possibilities:

Perhaps neither of the above formulae will survive further tests of other delicate observations. In this event we are not at the end of our resources. There are, in addition to Einstein's, yet two other sets of tensor differential equations which on the theory of nature explained in this lecture satisfy all the general requirements. These requirements are, (i) to have no arbitrary reference to any one particular time-system, and (ii) to give the Newtonian term of the inverse square law, and (iii) to yield the small corrections which explain various residual results which cannot be deduced as effects of the main Newtonian law.

The possibility of other such laws, expressed in sets of differential equations other than Einstein's, arises from the fact that on my theory there is a relevant fact of nature which is absent on Einstein's theory. This fact is the whole bundle of alternative time stratifications arising from the uniform significance of events [R, pp. 84–85].

In order to formulate these two other laws of gravitation, Whitehead makes use of the tensors $\|J_{\mu\nu}\|$ and $\|G_{\mu\nu}\|$ (see R, pp. 85–86). In Cartesian coordinates the two laws take on the forms:

$$
\frac{1}{\omega_\rho^2} \frac{\partial}{\partial x_\rho} J\,[\mu\nu,\,\rho]^{(x)} = 0 \qquad (\mu,\nu = 1, 2, 3, 4)\,,
$$

and

$$
\frac{\partial}{\partial x} T^{\rho\sigma}_{(x)} J\,[\mu\nu,\,\sigma]^{(x)} = 0 \qquad (\mu,\nu = 1, 2, 3, 4)\,,
$$

where

$$
J\,[\mu\nu,\,\lambda]^{(x)} = \frac{1}{2} \left\{ \frac{\partial J^{(x)}_{\mu\lambda}}{\partial x_\nu} + \frac{\partial J^{(x)}_{\nu\lambda}}{\partial x_\mu} - \frac{\partial J^{(x)}_{\mu\nu}}{\partial x_\lambda} \right\} \quad (\lambda,\,\mu,\nu = 1, 2, 3, 4)\,,
$$

[19] The tensor $R^{\rho}_{\mu\nu\sigma}$, being of rank four, has 256 components; however, various symmetry conditions reduce these to just twenty independent components. See P. Bergmann, *Introduction to the Theory of Relativity*, pp. 172–74.

[20] Cf. A. S. Eddington, *The Mathematical Theory of Relativity*, p. 95: "The field described by the $g_{\mu\nu}$ may be (artificially) divided into a *field of pure inertia* represented by the Galilean values, and a *field of force* represented by the deviations of the $g_{\mu\nu}$ from the Galilean values."

and $\|T^{\mu\nu}_{(x)}\|$ is some contravariant tensor connected with the electromagnetic field. Whitehead does not investigate the consequences of the two above laws of gravitation, but the first of them has been studied by G. Temple (see Appendix IV).

6. Comparison of Einstein's and Whitehead's Theories of Relativity

I shall now attempt to summarize and amplify my earlier remarks on the fundamental differences between Einstein's and Whitehead's theories of relativity. The law of motion is essentially the same in the two theories; I shall therefore concentrate on the law of gravitation. *Mathematically*, there is an immense difference between the two laws of gravitation. Einstein's law (equations [8.45]) is expressed by a set of ten non-linear differential equations of the second order, involving $g_{\mu\nu}$ and its first and second derivatives with respect to the four coordinates. Whitehead's law, on the other hand, can be expressed most simply in the integral form (equation [32]), which provides an explicit expression for the tensor $\|J_{\mu\nu}\|$ of the gravitational field. To take the simplest possible case—motion in the field produced by a mass-point at rest: in Einstein's theory, one must solve a set of non-linear differential equations (with all the attendant mathematical difficulties of such a solution) in order to obtain the $g_{\mu\nu}$ which define the gravitational field, while in Whitehead's theory one *begins* with a relatively simple expression for the gravitational field. In either case, after finding the ten tensor-components for the gravitational field, one substitutes them in the equations of motion and solves for the required orbit. Here both theories are confronted with formidable mathematical difficulties even in the case of problems which are relatively easy to handle in terms of classical gravitational theory.

The above mathematical differences in Einstein's and Whitehead's theories of gravitation are closely connected with certain important *physical* differences. There are at present very few known applications of Einstein's field equations (by an "application" here is meant a *rigorous* solution of the field equations without simplifying approximations). Thus, Whitehead points out (R, p. 83) that the solution of Einstein's field equations for a single static mass-point (the Schwarzschild solution) is not applicable to the moon's motion, whereas Whitehead's law of gravitation is applicable to all such cases. Also, the entire apparatus of special relativity dynamics (described above, pp. 175–80) can obviously be incorporated within Whitehead's theory of relativity.

In some cases the failure to find a rigorous solution of Einstein's field equations may be due simply to the mathematical difficulties involved; on the other hand, in certain cases no solution may exist[21] and furthermore, in certain other

[21] Cf. *ibid.*, p. 95: "No solution of Einstein's equations has yet been found for a field with two singularities or particles. The simplest case to be examined would be that of two equal particles revolving in circular orbits round their centre of mass. Apparently there should exist

cases, it may even be impossible to formulate the problem in terms of the general theory of relativity.[22] There appears also to be a certain circularity in Einstein's general field equations (8.47): the tensor $T_{\mu\nu}$ must be defined in metrical terms and therefore presupposes definite values for the $g_{\mu\nu}$ in a given region of space-time; while the metrical tensor $g_{\mu\nu}$ (of which $R_{\mu\nu}$ is a function) is itself supposed to be determined by $T_{\mu\nu}$.[23]

Connected with both the physical and the mathematical differences in their respective theories of relativity are the opposed *philosophical* principles of Einstein and Whitehead. Two points in particular are singled out by Whitehead in his criticism of the philosophical inadequacies of Einstein's theory, namely, the inability of that theory to give coherent accounts of measurement and of rotation. In each case Whitehead holds that Einstein's theory is inconsistent both with our direct perceptual knowledge of nature and with the actual working procedures of scientists. First, as to measurement:

By identifying the potential mass impetus of a kinematic element with a spatio-temporal measurement Einstein, in my opinion, leaves the whole antecedent theory of measurement in confusion, when it is confronted with the actual conditions of our perceptual knowledge. The potential impetus shares in the contingency of appearance. It therefore follows that measurement on his theory lacks systematic uniformity and requires a knowledge of the actual contingent physical field before it is possible. For example, we could not say how far the image of a luminous object lies behind a looking-glass without knowing what is actually behind that looking-glass [R, p. 83].

---

a statical solution with two equal singularities; but the conditions at infinity would differ from those adopted for a single particle since the axes corresponding to the static solution constitute what is called a rotating system. The solution has not been found, and it is even possible that no such statical solution exists. . . .

"The problem of two bodies on Einstein's theory remains an outstanding challenge to mathematicians—like the problem of three bodies on Newton's theory.

"For practical purposes methods of approximation will suffice. We shall consider the problem of the field due to the combined attractions of the earth and sun, and apply it to find the modifications of the moon's orbit required by the new law of gravitation. The problem has been treated in considerable detail by de Sitter."

[22] See Synge, "Orbits and Rays in the Gravitational Field of a Finite Sphere According to the Theory of A. N. Whitehead," p. 307: ". . . the number of problems which can be completely *formulated* (let alone solved) in the general theory of relativity is very small indeed. The simple geophysical problem of determining the change in period of a pendulum due to local variation in the earth's density, for example, seems impossible to formulate in terms of the general theory of relativity, and the problem of the tides presents the same difficulty."

[23] Another problematic aspect of the $g_{\mu\nu}$ in Einstein's theory is suggested by Synge: ". . . in Einstein's theory there is no space-time defined topologically which in [*sic;* should read "in which"] the g_{mn} are functions of position, and the g_{mn} seem to have then the responsibility of determining the topology. The assumption that there is no measure of separation in space-time except $g_{mn}dx_mdx_n$ leads to some oddities, for if a particle is a singularity at which some of the g_{mn} become infinite, it may become inaccessible through the infinite length of world lines drawn to it. This sort of thing cannot happen in Whitehead's theory, because the gravitational field is displayed against a flat 4-space with Euclidean topology, and an infinity in the g_{mn} causes no embarrassment at all (anymore than does an infinite potential in Newtonian gravitation)" (*The Relativity Theory of A. N. Whitehead,* p. 10).

And again:

> I cannot understand what meaning can be assigned to the distance of the sun from Sirius if the very nature of space depends upon casual intervening objects which we know nothing about. Unless we start with some knowledge of a systematically related structure of space-time we are dependent upon the contingent relations of bodies which we have not examined and cannot prejudge [R, pp. 58–59].

We recall that perception in the mode of relatedness assures us that space-time must be uniform; now Whitehead argues that such uniformity is presupposed in all discussions by scientists of hypothetical distances between real or hypothetical objects.

The problem of rotational motion has, of course, been of decisive importance at many points in the history of mechanics. Newton, for example, believed that absolute motion could most readily be detected in the case of rotations, whose *causes* are observable centripetal forces and whose *effects* are observable centrifugal forces. In Einstein's special theory of relativity rotational motions remain absolute (in spite of the fact that Einstein gives up Newton's absolute space and absolute time). In fact, one of Einstein's reasons for seeking an extension of the postulate of special relativity was precisely the necessity of an epistemologically satisfactory account of rotational motion. Thus, toward the beginning of his "The Foundation of the General Theory of Relativity"[24] Einstein introduces the following thought-experiment: We imagine two identical fluid bodies hovering freely in space at such great distances from one another and from all other bodies that no gravitational forces need be considered other than those arising from the interaction of the various parts of the same body. Suppose now that the distance between the two bodies is constant and that there is no relative motion of the parts of either body. Finally, let there be a relative rotational motion of constant angular velocity about the line joining the two bodies. Now, imagine that the surface of each body has been measured with instruments at rest relative to the body, and suppose one surface turns out to be that of a sphere and the other that of an ellipsoid. We ask: Why do the surfaces of the two bodies differ? Newton's answer to this question would have been that the laws of mechanics hold only for the coordinate-system in which the spherical body is at rest, the ellipsoidal shape of the other body being produced by its absolute rotation with respect to all Galilean coordinate-systems. This answer Einstein rejects on the ground that the privileged Galilean coordinate-systems are "factitious," or unobservable causes, and "No answer can be admitted as epistemologically satisfactory, unless the reason given is an *observable fact of experience.*"[25] Einstein concludes that the only satisfactory answer is that the cause of the differences in the two bodies must lie *outside* of the system consisting solely of the two bodies—presumably in the differential influence of the positions and motions of distant bodies relative to the sphere and ellipsoid.

[24] See H. A. Lorentz *et al.*, *The Principle of Relativity*, p. 112.

[25] *Ibid.*, pp. 112–13.

Generalizing his conclusion from this example, Einstein arrives at his principle that there can be *no* privileged coordinate-systems for the expression of physical laws.[26]

Whitehead, on the contrary, insists that the Galilean coordinate-systems *do* have a privileged status among all possible coordinate-systems because in terms of these special coordinate-systems one can give simple, intuitively plausible explanations of such rotational phenomena as Foucault's pendulum, the equatorial bulge of the earth, the fixed sense of rotation of cyclones and anti-cyclones, and the operation of the gyro-compass. (For Whitehead's discussion and critique of the classical prerelativistic explanation of rotation in terms of absolute space and a material ether, see PNK, pp. 34 ff.) Whitehead's interpretation of rotational motion is explained in the following passage:

The effects of rotation are among the most widespread phenomena of the apparent world, exemplified in the most gigantic nebulae and in the minutest molecules. The most obvious fact about rotational effects are their apparent disconnections from outlying phenomena. . . . The Einstein theory in explaining gravitation has made rotation an entire mystery. Is the earth's relation to the stars the reason why it bulges at the equator? Are we to understand that if there were a larger proportion of run-away stars, the earth's polar and equatorial axes would be equal, and that the nebulae would lose their spiral form, and that the influence of the earth's rotation on meteorology would cease? Is it the influence of the stars which prevents the earth from falling into the sun? The theory of space and time given in this lecture, with its fundamental insistence on the bundle of time-systems with their permanent spaces, provides the necessary dynamical axes and thus accounts for these fundamental phenomena. I hold this fact to be a strong argument in its favour, based entirely on the direct results of experience [R, pp. 87–88].

Whitehead nowhere elucidates the way in which his theory of relativity accounts for rotational phenomena, but we have already learned (see above, p. 180) how the forces associated with rotation may be introduced within the dynamical framework of special relativity (which is essentially the dynamical framework of Whitehead's theory of relativity). Whitehead argues that any adequate explanation of rotational effects must be independent of "outlying phenomena"—which means, of course, that centrifugal forces must be interpreted as fictitious in character since they are directly proportional to the distance from the axis of rotation.[27]

According to Einstein's general theory of relativity, on the other hand, "inertial" forces (such as those associated with accelerated coordinate-systems) must be identified with "gravitational" forces. Thus, Einstein illustrates his principle of equivalence by the example of a freely falling box—precisely the

[26] Cf., however, on the failure of general relativity itself to eliminate entirely the concept of absolute space, A. Grünbaum, "The Philosophical Retention of Absolute Space in Einstein's General Theory of Relativity."

[27] For a rejoinder to Whitehead's type of criticism of general relativity, see Eddington, *The Mathematical Theory of Relativity*, pp. 99–100.

same example used above to illustrate inertial forces (see above, pp. 168–69). The idea that inertia is due (like gravitation) to the *interaction* of material bodies—rather than being, as in the Newtonian view, an *intrinsic* property of matter—goes back to Mach. Einstein, therefore, designates as "Mach's principle" the general hypothesis that the character of space-time is determined by the distribution of matter and energy; this accounts for Einstein's dictum that "There can be no space nor any part of space without gravitational potentials; for these confer upon space its metrical qualities, without which it cannot be imagined at all."[28] Hence, Einstein considers it to be a great advantage of his general theory of relativity that it predicts the following three effects (none of them large enough to be experimentally confirmed at the present time): (1) the inertia of a body will increase as ponderable masses accumulate in its neighborhood; (2) a body will experience an accelerating force when neighboring masses are accelerated; (3) a rotating hollow body will generate inside of itself both a Coriolis field and a radial centrifugal field of forces.[29] In summary, one might say Einstein supposes that matter (or the gravitational field) is ontologically prior to space-time, while events are simply intersections of world-lines of particles; and Whitehead supposes that events are ontologically prior to space-time, while matter is simply a contingent characteristic of certain events.[30]

[28] A. Einstein, "Ether and Relativity," *Sidelights on Relativity*, p. 21. For a recent discussion of the—rather problematic—current status of Mach's principle, see Grünbaum, "The Philosophical Retention of Absolute Space in Einstein's General Theory of Relativity."

[29] See A. Einstein, *The Meaning of Relativity*, p. 100; and C. Møller, *The Theory of Relativity*, p. 317.

[30] It is interesting to note that when Einstein visited England in 1921 he and Whitehead discussed the theory of relativity: "Whitehead had long discussions with Einstein and repeatedly attempted to convince him that on metaphysical grounds the attempt must be made to get along without the assumption of a curvature of space. Einstein, however, was not inclined to give up a theory, against which neither logical nor experimental reasons could be cited, nor considerations of simplicity and beauty. Whitehead's metaphysics did not seem quite plausible to him." (P. Frank, *Einstein, His Life and Times*, p. 189.) I am indebted to Professor A. J. Coleman of the Mathematics Department of the University of Toronto for this reference. (Cf. also F. S. C. Northrop's "Whitehead's Philosophy of Science," *The Philosophy of Alfred North Whitehead*, ed. P. A. Schilpp, p. 204, for a report of a conversation with Einstein on Whitehead's conception of simultaneity.)

CHAPTER X

CONCLUSION: NATURAL SCIENCE AND THE PHILOSOPHY OF ORGANISM

By far the most important influence on Whitehead's philosophy of science, as has been abundantly illustrated in the preceding chapters, was Einstein's theory of relativity. More difficult to assess is the influence on Whitehead's philosophy of organism of the other great revolutionary development in twentieth-century physics, namely, quantum theory. It is impossible—but fortunately also unnecessary—to give a comprehensive account of quantum theory in this final chapter. However, a bit of chronology will be helpful.[1] It was in 1900 that Planck introduced the quantum idea into physics when he produced the first adequate explanation of black-body radiation by postulating that only certain discrete energies were possible for radiation of any given frequency. In 1905 Einstein proposed an explanation of the photoelectric effect by postulating that light can be absorbed and emitted by atoms only in discrete and indivisible quantities, called "photons," which behave in certain respects like tiny corpuscles of matter moving with the speed of light. Both Planck's and Einstein's ideas were incorporated in Bohr's theory of the atom (1913). The crucial quantum postulates of Bohr were that (1) the electrons revolving about the central nucleus are restricted to (periodic) orbits with certain discrete values for their dimensions, angular momenta, and energies; and (2) light is absorbed or emitted by atoms in the form of photons when an electron "jumps" from one orbit to another. Between 1916 and 1926 Sommerfeld, Born, Jordan, and others refined and extended Bohr's theory. In 1924 de Broglie predicted on the basis of abstract theoretical considerations that elementary particles (such as electrons) should exhibit wave-like characteristics under certain conditions, and in 1925 Davisson and Germer and independently G. P. Thomson confirmed this prediction experimentally by showing that electrons could be diffracted like

[1] An excellent untechnical account of the development of quantum theory may be found in L. de Broglie, *The Revolution in Physics*.

light-waves. Later experiments led to the discovery of wave-like characteristics in other types of particles: protons, neutrons, and alpha-particles; and, with less certainty, atoms and molecules.

Each of the theories just mentioned could be (and was) criticized on two counts: first, the quantum conditions seemed to be *ad hoc* postulates, introduced without any convincing theoretical justification; and, second, none of the theories was truly comprehensive (e.g., the Bohr-Sommerfeld model of the atom worked well only for the hydrogen atom). Beginning in 1925–26 with the work of Born, Heisenberg, Schrödinger, Dirac, Jordan, and Pauli, a theory— or better, a group of related theories—was developed which met the two above criticisms: quantum mechanics in any of its various formulations (matrix mechanics, wave mechanics, etc.) has succeeded in deducing from a small set of general and more or less plausible assumptions all the quantum conditions needed to explain an astonishingly broad range of phenomena (including, e.g., in principle and increasingly in practice, the nature of chemical bonds). On the other hand, it is exceedingly difficult to find an intuitive picture or model for quantum mechanical explanations—at least, this is the case if one is reluctant to abandon traditional scientific conceptions of the nature of material particles (and "traditional" here is meant to embrace relativistic as well as classical conceptions). Hence the uneasiness of some physicists (notably Einstein and de Broglie) over the apparent implications of quantum mechanics for the nature of physical reality, an uneasiness which takes the form of a refusal to accept as ultimate theoretical principles the basic assumptions of quantum mechanics (such as Heisenberg's Uncertainty Principle). Quantum mechanics, as originally formulated, was concerned primarily with electrons, either in the free state or inside atoms (atomic nuclei came in only insofar as they affected the electrons inside atoms), and with the interaction of photons and electrons (both within and outside of atoms). After 1930 the discovery of new elementary particles (neutrons, positrons, mesons, etc.) demanded further theoretical advances to account for nuclear forces and for the production and annihilation of elementary particles; elementary particle theory is not yet today in a completely satisfactory state. Finally, the problem of unifying relativity theory and quantum theory should be mentioned. Special relativity has come to play a central role in quantum mechanics (e.g., Dirac's Lorentz-covariant reformulation of quantum mechanics led to the prediction of the particle now called the positron). The relation between general relativity and quantum mechanics, on the other hand, has been studied very little until recently, and there is even some question as to the compatibility of the two theories.

The chronology of the above summary of the development of quantum theory demonstrates that only Planck's, Einstein's, and Bohr's contributions could possibly have influenced the original formulation of Whitehead's philosophy of organism, which dates from 1925—the year in which Whitehead delivered the Lowell Lectures (subsequently published as *Science and the Modern World*). The internal evidence of Whitehead's writings suggests, in fact, that he never became acquainted with the post-1924 developments in quantum theory or perhaps even with de Broglie's "matter waves," since there is not a single allusion to any of these developments or to their authors in all of Whitehead's published works. I shall return to this somewhat puzzling fact below.

The reformulation and reinterpretation which Whitehead's theory of extension undergoes within the context of the philosophy of organism are paralleled by corresponding changes in Whitehead's theory of objects. In neither case do the changes constitute simple contradictions of any of the major doctrines of Whitehead's natural philosophy; rather, certain implicit ideas of that natural philosophy become explicit and the discussion proceeds at an entirely new—ontological—level. For our purposes the principal changes in the theory of objects may be epitomized by the shift in the meanings of scientific objects and of percipient objects.

In the earlier theory of objects, sense-objects are primary and all other types of objects are derivative. Thus scientific objects (e.g., electrons or atoms) are construed as aspects or properties of physical objects, which must themselves be defined in terms of sense-objects. Scientific objects may be uniform (as in the case of electrons and protons) or non-uniform (as in the case of atoms and molecules). Percipient (or "living") objects really transcend the categories of natural philosophy but they are briefly discussed under the less misleading term "rhythms," where a rhythm is a unique association of a complex object, or pattern, and a sequence of events (see above, p. 164). Any non-uniform object necessarily involves some rhythm, and to this extent there is an analogy between living organisms and at least some of the fundamental scientific objects. (One of the first and most significant steps in Whitehead's development of his philosophy of organism is to broaden and deepen this analogy.)

In the philosophy of organism the primary objects—now called "eternal objects"—belong to one of two species: subjective forms of feeling (at least some sensa belong to this category) and objective mathematical forms (at least the concepts of the theory of extension belong to this category); while electrons, atoms, and living things, as we shall see, all belong to a single ontological category, namely, all are "societies" of one kind or another.

Before introducing the concept of a society we must examine Whitehead's attempt to assimilate all scientific objects to the non-uniform type. First of all, when Whitehead says that atoms and molecules are non-uniform it is fairly evident that he is thinking of the Bohr (or Bohr-Sommerfeld) theory—writing in the years between 1919 and 1925, Whitehead could have been thinking of scarcely any other theory. Hence Whitehead's attribution of non-uniformity to atoms must refer to the periodic character of the orbits of the electrons in the Bohr model of the atom (this periodicity being related to the periodicity of the radiation which can be emitted or absorbed by the atom).[2] As Whitehead

---

[2] The mathematical relation between the orbital periods (or frequencies) and the radiation periods (or frequencies) is purely formal and without direct physical significance except in the

interprets the Bohr model, then, an atom, like the waves of radiation which it can emit or absorb, is non-uniform in the sense that it requires some finite interval of time (equal presumably to the maximum orbital period of its electrons) in which to manifest its own characteristic properties.

What now of electrons? In 1920 we find Whitehead asserting the bare possibility of their non-uniformity on the ground that "Some such postulate is apparently indicated by the modern quantum theory" (CN, p. 162); and in 1925 he suggests:

> The discontinuities introduced by the quantum theory require revision of physical concepts in order to meet them. In particular, it has been pointed out that some theory of discontinuous existence is required. What is asked from such a theory, is that an orbit of an electron can be regarded as a series of detached positions, and not as a continuous line [SMW, p. 196].

Assuming once more that "quantum theory" here means Bohr's quantum theory, I take it that Whitehead is referring to the difficulty in accounting for the process by which an electron makes its transition from one orbit to another—a process concerning which Bohr's theory furnishes no information whatsoever. Whitehead proposes that a way out of this difficulty might be to develop a theory in which the electron is conceived as "a vibratory ebb and flow of an underlying energy, or activity" (SMW, p. 53).[3] Whitehead explains what such a theory might be like and how it could be taken as an exemplification of the main categories of the philosophy of organism (see SMW, chap. viii). Thus, two essentially different types of vibrations are distinguished: vibratory *locomotion* of a given pattern as a whole, whose laws are those of relativity theory; and vibratory *organic deformation* (or change of pattern), whose laws are those of quantum theory. The entities of such a theory will be *organisms* so that "the plan of the *whole* influences the very characters of the various subordinate organisms which enter into it" (SMW, p. 115). The ultimate organisms are those not decomposable into other organisms; these Whitehead calls *primates* and tentatively identifies with electrons and protons. Also, he is careful to point out that physics is concerned not with the properties of primates in themselves but rather with those properties of primates whose "effects on patterns and on locomotion are expressible in spatio-temporal terms" (SMW, p. 191). Perhaps, then, Whitehead speculates, when a primate changes its proper

---

limiting case of very large orbits when—in accordance with Bohr's Correspondence Principle—the frequency of the radiation emitted or absorbed as an electron jumps between two successive orbits is equal to the average of the two orbital frequencies of the electron.

[3] We find Whitehead repeating essentially the same argument in his Gifford Lectures of 1927–28: ". . . the quanta of energy are associated by a simple law with the periodic rhythms which we detect in the molecules. Thus the quanta are, themselves, in their own nature, somehow vibratory; but they emanate from the protons and electrons. Thus there is every reason to believe that rhythmic periods cannot be dissociated from the protonic and electronic entities" (PR, p. 122).

spatio-temporal systems very rapidly (or, is rapidly accelerated), it is deformed beyond recovery, i.e., "goes to pieces" and "dissolves" into radiation of the same period. The *endurance* of a primate thus depends on the existence of a favorable environment, exactly as in the case of living organisms. This general principle of endurance or survival Whitehead arrives at by generalizing what he takes to be a fundamental principle of the theory of evolution.[4] Indeed, the philosophy of organism seems to owe at least as much to biology (and to sociology) as to physics. But, at a deeper level, modern biology and physics are converging in their fundamental concepts: "Science is taking on a new aspect which is neither purely physical, nor purely biological. It is becoming the study of organisms. Biology is the study of the larger organisms; whereas physics is the study of the smaller organisms" (SMW, p. 150).

Whitehead is well aware that his discussion of primates is no substitute for a physical theory expressed in mathematical form; and it would surely be unwarranted to hold that Whitehead had anticipated de Broglie's conception of the electron.[5] (It is of some interest, however, that de Broglie's original heuristic arguments were based on speculative analogies—though of a physical rather than a philosophical character.) As mentioned above, Whitehead never refers in unmistakable fashion to de Broglie's matter waves or to the later formulations of quantum mechanics. In view of the great intrinsic philosophical interest of the new quantum mechanics, Whitehead's utter silence on the subject is both astonishing and disappointing—especially disappointing since there does seem to be some general consonance between the broader implications of this theory and the categories of the philosophy of organism. However, it is idle to speculate on just how Whitehead would have interpreted the new quantum mechanics if he had studied it (which I suspect he did not).

One of the most important derivative notions of the philosophy of organism

[4] ". . . the key to the mechanism of evolution is the necessity for the evolution of a favourable environment, conjointly with the evolution of any specific type of enduring organisms of great permanence. Any physical object which by its influence deteriorates its environment, commits suicide" (SMW, pp. 160–61).

[5] Most important, of course, was de Broglie's expression for the wave-length associated with moving particles, namely, $\lambda = h/p$, where λ is wave-length, h is Planck's constant, and p is momentum. This expression made possible the testing of de Broglie's hypothesis by diffraction experiments. Incidentally, another anticipation of an important physical theory sometimes attributed to Whitehead is even further off the mark: it is alleged that Whitehead's conception of the electron as extending throughout all space foreshadows Schrödinger's wave-mechanical conception of the electron. Not at all. Whitehead, it will be recalled (see above, p. 155), distinguishes the occupied from the unoccupied events in the field of an electron; whereas one of the most remarkable features of Schrödinger's conception is precisely that, in certain (not uncommon) circumstances, this distinction breaks down, i.e., the electron must literally be imagined as spread out through all of space. (Is it to Schrödinger's conception that Whitehead is referring when he says that "the latest speculations tend to remove the sharp distinction between the 'occupied' portions of the field and the 'unoccupied' portion" [PR, p. 112]?) Incidentally, Born's alternative interpretation of Schrödinger's wave mechanics—favored today by the majority of physicists—in terms of the probability of finding an electron at a given point hardly brings wave mechanics any closer to Whitehead's conception of the electron.

is that of a nexus with "social order," or a "society." This notion can be defined in terms of the categoreal scheme of the philosophy of organism (see PR, pp. 50–51, and AI, p. 261); in particular, the definition makes use of four of the categories of existence: actual occasions, eternal objects, prehensions, and nexūs (all of which have already been encountered in another context—see nn. 8, 10, 11, and 16, chap. vi). Roughly speaking, a *society* is a nexus of actual occasions, all of them characterized by a certain complex eternal object (called the *defining characteristic* of the society), which is inherited by later from earlier members of the nexus. A nexus of mutually contemporary occasions cannot constitute a society because inheritance is impossible in such a nexus.[6] On the other hand, societies with purely temporal and continuous (or serial) order are peculiarly simple and important; such societies Whitehead terms *persons* or *enduring objects*. A human being conceived of as a single stream of memories is an enduring object in this sense (but a concrete human being is far more than a stream of memories—his body cannot be overlooked). Enduring objects need not be "living," e.g., the corpuscles of physics (including electrons, protons, photons, etc.) are—at least in first approximation—enduring objects; conversely, "living" societies need not be personal, e.g., neither a tree nor, Whitehead conjectures (see PR, p. 158), a single cell possesses personal order. A society analyzable into strands of enduring objects is termed *corpuscular* (e.g., ordinary material objects). Also, this corpuscular property is a matter of degree (Whitehead's example is a train of light-waves at various stages of its career).

There are no societies in isolation; each society implies a wider background, or environment, which together with the society forms a more inclusive society. Any social environment is dominated by a set of causal laws which arise from the defining characteristic of that society. It is important to note that a "society is only efficient through its individual members. Thus in a society, the members can only exist by reason of the laws which dominate the society, and the laws only come into being by reason of the analogous characters of the members of the society" (PR, p. 139). The universe may be described as a succession of cosmic epochs in each of which ". . . a system of 'laws' determining reproduction in some portion of the universe gradually rises into dominance; it has its stage of endurance, and passes out of existence with the decay of the society from which it emanates" (PR, p. 139).

To do justice to the full complexity of the societies within our present cosmic epoch Whitehead finds it necessary to introduce the idea of a "structured society." A *structured society* is one which includes subordinate societies and subordinate nexūs with a definite pattern of interrelations (PR, p. 151). The great advantage of a highly complex structured society is that it can intensify the satisfaction of some of its component members. (The ultimate source of the quest for intensity of satisfaction is God conceived as the "lure for feeling, the eternal urge of desire" [PR, p. 522].) Among the important types of structured

---

[6] In other words, it is essential to a society that it *endure:* "The real actual things that endure are all societies. They are not actual occasions" (AI, p. 262).

societies are molecules, crystals, and "in all probability" free electrons and protons (PR, p. 152). Thus once more, as in his earlier philosophy of science, Whitehead is not satisfied with his initial account of electrons and protons as ultimately simple entities (the primates of SMW, the corpuscles or enduring objects of PR), and he proceeds to qualify that account by suggesting that each of the so-called elementary particles of physics may be a different species of structured society. Elementary particles, atoms, molecules, and material bodies in general, represent one of two principal solutions provided by structured societies to nature's great problem, namely, to produce societies which unite a high degree of survival power with a complex structure (and hence the possibility of evoking intensities of feeling). It is obvious that, in general, there will be an inverse relation between these two attributes: a complex society will usually require a specialized environment to maintain it; while a relatively simple society will be able to adapt its structural pattern easily to the immediate circumstances in which it finds itself. This problem has been solved along two divergent lines. At one pole, there has been a gradual elimination of differences of detail in the members of certain societies, but, at the same time, the intensities of certain feelings in the societies is increased by "the massive objectifications of the many environmental nexūs, each in its unity as *one* nexus, and not in its multiplicity as *many* actual occasions" (PR, p. 154). At the other pole, there are societies whose members actively initiate conceptual prehensions: the subjective aim of each concrescent occasion originates novelty to match the novelty of the environment. Structured societies in which only the first mode of solution is important are termed *inorganic;* structured societies in which the second mode becomes important are termed *living*. Also, the second mode of solution always presupposes the first, since no living societies are known without a "subservient apparatus of inorganic societies" which exercises a protective and stabilizing function.

There is, clearly, something paradoxical about the above explanation of living societies in terms of novelty, for novelty can scarcely be a *defining characteristic* of anything. And guided apparently by an awareness of this paradox, Whitehead develops in greater detail his explanation of living societies. First of all, he notes that in a living society only certain nexūs will consist of actual occasions *each* of which introduces novel reactions in its concrescence. Nexūs of this type are called *entirely living*, and a society is called living only when it is dominated or controlled by such an entirely living nexus. But, says Whitehead, it is plausible to suppose that entirely living nexūs are not societies: ". . . in abstraction from its animal body an 'entirely living' nexus is not properly a society at all, since 'life' cannot be a defining characteristic. It is the name for originality, and not for tradition" (PR, pp. 159–60). Entirely living nexūs are somewhat deficient socially and, thereby, relatively more independent of the past and more capable of spontaneity than a society could be. And Whitehead concludes that such entirely living nexūs must be located in "empty" space, i.e., in space free of all inorganic nexūs and societies. "Life lurks in the interstices of each living cell, and in the interstices of the brain" (PR, p. 161). Furthermore, the transmission of physical influences through these interstices will not take place in accordance with the usual laws which hold for inorganic societies. "The molecules within an animal body exhibit certain peculiarities of behaviour not to be detected outside an animal body" (PR, p. 162). There is no implication, however, that the laws governing the behavior of molecules within an animal body are either non-existent or undiscoverable; or, in other words, that biology is anything more than a special branch of natural science. This brings us to the question of the scientific status of biology from the standpoint of the philosophy of organism.

Whitehead distinguishes two branches of physiology: physical and psychological. The former studies the "subservient inorganic apparatus" (i.e., the physical body) of a living organism and corresponds pretty much to what is called "physiology" today. The latter "seeks to deal with 'entirely living' nexūs, partly in abstraction from the inorganic apparatus, and partly in respect to their response to the inorganic apparatus, and partly in regard to their response to each other" (PR, pp. 157–58). According to Whitehead, psychological physiology is still in a state of infancy. One may be permitted to wonder how the science can ever grow up if it must deal with such an intractable subject matter as the spontaneities, the novelties, characteristic of the occasions of an entirely living nexus. Certainly, one should not expect in psychological physiology the rigor of physical or even that of ordinary biological laws. Perhaps psychological physiology should investigate the ways in which the "purposiveness" characteristic of a living thing as an individual whole is related to the "purposiveness" characteristic of the successive behavioral manifestations (including acts of perception) of that living thing. Or, translating into Whiteheadian terms, psychological physiology should investigate the interrelations of organisms in the microscopic and macroscopic senses:

The microscopic meaning is concerned with the formal constitution of an actual occasion, considered as a process of realizing an individual unity of experience. The macroscopic meaning is concerned with the givenness of the actual world, considered as the stubborn fact which at once limits and provides opportunity for the actual occasion. . . . in our experience, we essentially arise out of our bodies which are the stubborn facts of the immediate relevant past [PR, pp. 196–97].

The purposiveness associated with microscopic organisms is what Whitehead calls "subjective aim,"[7] while the purposiveness associated with macroscopic organisms is what Whitehead calls "physical purpose."[8] Also, I believe that

---

[7] "The 'subjective aim,' which controls the becoming of a subject, is that subject feeling a proposition with the subjective form of purpose to realize it in that process of self-creation" (PR, p. 37).

[8] Whitehead's explanation of physical purposes is too complicated to reproduce here (see PR, pp. 420–28); however, the cosmological role of such purposes is clear: "The constancy of physical purposes explains the persistence of the order of nature, and in particular of 'enduring objects'" (PR, p. 421). Also, physical purposes account for the association of endurance with rhythm and physical vibration.

physical purpose and subjective aim correspond closely to the two types of causation recognized by the philosophy of organism, namely, "efficient causation [which] expresses the transition from actual entity to actual entity; and final causation [which] expresses the internal process whereby the actual entity becomes itself" (PR, p. 228).[9]

Physical science can, for the most part, safely ignore final causation, since such causation is (though perhaps not totally absent) practically ineffective in those aspects of actual occasions of interest to physical science. Nevertheless, physics "In order to reconsider its foundations . . . must recur to a more concrete view of the character of real things, and must conceive its fundamental notions as abstractions derived from this direct intuition. It is in this way that it surveys the general possibilities of revision which are open to it" (SMW, p. 196). Thus philosophy—which for Whitehead always involves, above all, a return to the concreteness of experience—furthers the progress of science by helping it to envisage possibilities in the way of concepts and theories which might otherwise be overlooked. And, of course, philosophy must never repudiate science: "It has been a defect in the modern philosophies that they throw no light whatever on any scientific principles" (PR, p. 178). In particular, if a dualism of nature and human experience is to be avoided—and Whitehead considers this a primary objective in any ultimately satisfactory philosophical system—then we must "point out the identical elements connecting human experience with physical science" (AI, p. 237). Thus, for example, "The notion of physical energy, which is at the base of physics, must then be conceived as an abstraction from the complex energy, emotional and purposeful, inherent in the subjective form of the final synthesis in which each occasion completes itself" (AI, p. 239; cf. PR, p. 177).

We have come back to the theme which introduced this book: the proper interrelation of science and philosophy. How can Whitehead help us here? The ultimate value of Whitehead's major theoretical constructions—his philosophy of science, his theory of relativity, his philosophy of organism—cannot be foreseen, but the example of his sustained and serious attempt to relate science and philosophy, both to one another and to man's other chief concerns, can perhaps provide the kind of stimulus and guide so sorely needed in our present intellectual predicaments. The very effort to understand Whitehead's writings must at the least serve to enlarge the scope of our intellectual interests, to correct our intellectual provincialisms, and to remind us that both breadth of view and concentration on detail are essential to the highest intellectual achievements.

The full solemnity of the world arises from the sense of positive achievement within the finite, combined with the sense of modes of infinitude stretching beyond each finite fact [MT, p. 108].

[9] For an interesting attempt to apply some of the categories of the philosophy of organism to biological phenomena, see W. E. Agar, *A Contribution to the Theory of the Living Organism.* Another book by a biologist showing (and acknowledging) a considerable indebtedness to the philosophy of organism is R. S. Lillie's *General Biology and Philosophy of Organism.*

PRIMES AND

ANTIPRIMES IN

THE PRINCIPLES OF

NATURAL KNOWLEDGE,

THE CONCEPT OF NATURE,

AND *PROCESS AND REALITY*

The discussion of the concepts of primeness and antiprimeness in the text (pp. 51 ff.) follows PNK; later (in CN, pp. 87–89), Whitehead defines the two concepts somewhat differently. This latter treatment—although not as careful or detailed as the former—is worth summarizing here because Whitehead seems to have adopted it in his important formulation of the method of extensive abstraction in PR. The CN definitions of σ-primes and σ-antiprimes run as follows. An abstractive class is a *σ-prime* when, for some given condition σ, (1) it satisfies the condition σ, and (2) it is covered by every abstractive class which both is covered by it and satisfies the condition σ. An abstractive class is a *σ-antiprime* when, for some given condition σ, (1) it satisfies the condition σ, and (2) it covers every abstractive class which both covers it and satisfies the condition σ.

According to the CN definition, a σ-prime is covered, not—as in the PNK definition—by *all* abstractive classes satisfying σ, but only by those abstractive classes which satisfy σ *and* are covered by that σ-prime; and an analogous contrast may be drawn between the two definitions of σ-antiprimes. The CN definitions are clearly weaker than the corresponding PNK definitions; thus there are conditions σ which define σ-primes (or σ-antiprimes) in the CN sense but not in the PNK sense. For example, the condition σ = "converging to a point" defines a set of σ-primes in the CN sense, namely, the set of *all* abstractive classes which converge to points; while this condition σ does *not* define a set of σ-primes in the PNK sense, because obviously not all abstractive classes converging to points cover one another. In general, from a condition σ which defines a set of σ-primes in the CN sense, one can always derive another condition σ' which defines a set of σ'-primes in the PNK sense, as follows. Suppose the abstractive class a is a σ-prime in the CN sense; this means that (1) a satisfies σ, and (2) for every abstractive class β, if β satisfies σ and a covers β, then β covers a. Now, let $\sigma' =$ "σ and being covered by a"; (1) and (2) become: (1') a satisfies σ' (since a covers itself), and (2') for every abstractive class β, if β satisfies σ', then β covers a. Thus, a is a σ'-prime in the PNK sense. Since σ' will, in general, designate different conditions for different choices of a (say, a, a', a'', etc.), what the definition of σ' has in effect done is to subdivide the class of σ-primes into σ_a'-primes, σ_a''-primes, σ_a'''-primes, etc. Thus, in our earlier example, the σ_a'-primes might correspond to the set of abstractive classes converging to one particular point, the σ_a''-primes to the set of abstractive classes converging to some other point, and so on.

It should by now be clear that one of the essential characteristics of σ-primes (or σ-antiprimes) in the PNK sense is absent from σ-primes (or σ-antiprimes) in the CN sense, namely, all σ-primes (or σ-antiprimes) in the former sense are K-equal, while in the latter sense this may or may not be the case (depending on the condition σ). Now, as a matter of fact it is fundamental to the method of extensive abstraction, in any of Whitehead's formulations, that each distinct geometrical entity (such as a particular point or a particular line) be defined in terms of a single complete set of K-equal abstractive classes. This explains why, in PNK, geometrical entities are always defined simply as sets of σ-primes (or σ-antiprimes), whereas in CN geometrical entities must (at least sometimes) be defined as sets of abstractive classes K-equal to a given σ-prime (or σ-antiprime). Furthermore, it cannot generally be proved that, for a given σ, an abstractive class K-equal to a σ-prime (or σ-antiprime) in the CN sense is itself a σ-prime (or σ-antiprime); or, put another way, it cannot generally be proved that, for a given σ, the set of abstractive classes K-equal to a given σ-prime (or σ-antiprime) is identical with the complete set of σ-primes (or σ-antiprimes). It was in order to avoid this difficulty that the concept of *regular* conditions for primes and antiprimes was introduced in PNK (see above, pp. 52–53). In CN, on the other hand, Whitehead is simply careful to choose his formative conditions for primes and antiprimes in such a way that the desired property can be proved (see, for example, the definition of a moment, p. 57 above).

In PR "prime" is defined just for members of a "geometrical element," i.e., for complete sets of equivalent (or K-equal) abstractive sets, thus avoiding the possibility of primes (with respect to assigned conditions) which are nonequivalent. That the PR definition of a prime abstractive set is essentially the same as the CN definition of a σ-prime may be seen as follows. First, we reformulate the PR definition (see PR, p. 457, Definition 16.1): An abstractive set a is a σ-*prime* when for some given condition σ, (1) a satisfies the condition σ, (2) every abstractive set equivalent to a also satisfies the condition σ, and (3) there is no abstractive set satisfying the condition σ and covered by a but not equivalent to a. Now, it is easily shown that (3) is logically equivalent to: every abstractive set which satisfies the condition σ and is covered by a also covers a. Thus, the PR definition, ignoring (2), is seen to be identical with the CN definition.

SOME GEOMETRICAL

CONCEPTS RELATED

TO WHITEHEAD'S

THEORY OF

EXTENSION

In his article "A Set of Postulates for Abstract Geometry, Expressed in Terms of the Simple Relation of Inclusion," E. V. Huntington presents a new set of postulates for ordinary three-dimensional Euclidean geometry. The principal novelty in this set of postulates lies in the choice of primitive concepts, of which there are just two: *sphere* and *contains*. (It turns out that all the postulates are true if the class of "spheres" is interpreted as the class of ordinary Euclidean spheres [including the null sphere] and if "contains" is interpreted in the ordinary spatial sense; but, of course, the two primitive terms need not be interpreted in this way nor need they be interpreted at all.)

Huntington begins with definitions for fundamental geometrical entities and relations such as point, straight line, plane, parallelism, congruence, and perpendicularity. Each of these definitions is formulated (ultimately) in terms of the two primitive concepts, e.g., a *point* is any sphere which contains no other sphere (Definition 4), and a (straight) *segment AB* is a class of points X such that every sphere which contains A and B also contains X (Definition 5), and a (straight) *line AB* is uniquely determined by the segment AB (Definition 8). (Huntington points out that "there is nothing in this definition [of a point], or in any of our work, which requires our 'points' to be *small;* for example, a perfectly good geometry is presented by the class of all ordinary spheres whose diameters are not less than one inch; the 'points' of this system are simply the inch-spheres" [pp. 529–30].)

Following the definitions comes the formulation of postulates which guarantee that the defined entities and relations will possess the usual properties of three-dimensional Euclidean geometry. It is assumed, for example, that the relation of containing is transitive (Postulate 1); that if two lines are parallel to a third line, they are either parallel or coincident (Postulate 9); and that if

AB is a line and C a point not on that line, then there is a point X such that CX is parallel to AB (Postulate E3). Most important for the purpose of comparing Huntington's system with Whitehead's theory of extension is Postulate 5: "If two lines have two distinct points in common, they coincide" (p. 537). The uniqueness characteristic of straight lines must be assumed in Huntington's system because it cannot be deduced from the definition of straight lines alone. Of course, if "sphere" is interpreted in its ordinary sense, then it easily follows from Huntington's definition of a (straight) segment that there is only one such segment between any two given points. Since, however, Huntington's system is not interpreted, the possibility of "curved" segments must be explicitly excluded. Postulate 5 (together with the other postulates exclusive of those concerning congruence) actually has the effect of restricting "spheres," not to ordinary Euclidean spheres but simply to the class of Euclidean *convex* three-dimensional solids. Later postulates, in particular those concerning congruence, do indeed restrict "spheres" to ordinary Euclidean spheres. (Incidentally, one of the most important parts of Huntington's article is devoted to a proof that his postulates are consistent, independent, and categorical.)

Among the advantages of his set of postulates for Euclidean geometry, according to Huntington, is "the fact that all *metric properties* are obtained directly in terms of the fundamental concepts, without the intervention of Cayley's 'absolute' " (p. 524). The shortcomings of Huntington's procedure from Whitehead's point of view would seem to be, first, that it assumes rather than deduces the uniqueness property characteristic of (straight) segments and, second, that it fails to formulate the general common aspects of Euclidean and non-Euclidean geometries. These points can be explained in more detail as follows. It will be recalled that Whitehead's theory of extension, as it is developed in PR, consists of assumptions and definitions, all of which are purely topological in character. The goal of this theory of extension is achieved when "points," "straight lines," and "planes" (in, essentially, the projective sense) have been defined and various properties of these entities have been deduced from their definitions. In particular, it is possible to deduce the following four properties of straight lines: (1) their completeness (the impossibility of adding "new" points to a given straight line), (2) their inclusion of points, (3) their unique definition by any pair of included points, and (4) the possibility of their intersection in a single point. Now, in order to compare Huntington's system with Whitehead's we must confine our attention to Huntington's definitions of point and straight line and to his postulates (1–5, E1 and E2) for points and straight lines. It is easy to show on the basis of these definitions and postulates that Huntington's straight lines do indeed possess the above four properties. However, the third— and for Whitehead the crucial—property is simply postulated by Huntington, whereas Whitehead deduces it from his definition of straight lines, essentially from the "oval" character of the regions which figure in that definition. Thus, Whitehead prefers to begin with rather complicated topological postulates and then to deduce the uniqueness condition for straight segments, while Huntington is content simply to postulate this condition at the outset. Furthermore, Huntington's later postulates immediately introduce the parallelism and congruence properties of Euclidean straight lines; whereas Whitehead prefers to take the route through projective geometry, thereby first treating explicitly the general (Euclidean or non-Euclidean) intersection properties of points and straight lines and then defining metrical concepts in a single unified way for all uniformly-curved spaces. On the other hand, though Huntington admits that his method is "not so readily adapted to the study of projective geometry" (p. 523), this formal disability must be shared by Whitehead's method since the intersection properties of points and straight lines are the same in both methods. (It will be recalled that Whitehead does not develop his theory of extension to the metrical stage.) The simple fact remains that Whitehead favors the intervention of Cayley's absolute while Huntington prefers to avoid it.

K. Menger has given an axiomatic development of topology which resembles both Huntington's and Whitehead's systems in that it takes finite regions rather than points as primitive entities. Furthermore, Menger's conception of the value of his approach is similar to Whitehead's: ". . . a topology of lumps seems to me to be closer to the physicist's concept of space than is the point set theoretical concept. For naturally all the physicist can measure and observe are pieces of space, and the individual points are merely given as the result of approximations" ("Topology without Points," p. 85). Also, the following passage implies a criticism of the method of extensive abstraction and a program for future developments of that method, which are very much in the spirit of Whitehead's natural philosophy.

In concluding, I should like to point out that even the introduction of points as nested sequences of lumps somehow transcends what can be observed in nature. For, by a lump, we mean something with a well defined boundary. But well defined boundaries are themselves results of limiting processes rather than objects of direct observation. Thus, instead of lumps, we might use at the start something still more vague— something perhaps which has various degrees of density or at least admits a gradual transition to its complement. Such a theory might be of use for wave mechanics ["Topology without Points," p. 107].

VECTORS

AND

TENSORS

In order to define the concept of a vector we consider transformations from one rectangular Cartesian coordinate-system to another. Such transformations, called "orthogonal transformations," have the form:

$$\left.\begin{aligned}
x' &= c_{11}x + c_{12}y + c_{13}z + x_0' \,, \\
y' &= c_{21}x + c_{22}y + c_{23}z + y_0' \,, \\
z' &= c_{31}x + c_{32}y + c_{33}z + z_0' \,,
\end{aligned}\right\} \quad (1)$$

where (x_0', y_0', z_0') represents the origin of the primed coordinate-system relative to the unprimed coordinate-system, and the c_{ij} are constant coefficients satisfying the conditions:

$$\sum_{i=1}^{3} c_{ij} c_{ik} = \delta_{jk} \left\{\begin{array}{l} = 0 \text{ for } j \neq k \\ = 1 \text{ for } j = k \end{array}\right\} \quad (j, k = 1, 2, 3) \,. \quad (2)$$

The transformation equations for the coordinate differences follow immediately from equations (1):

$$\left.\begin{aligned}
\Delta x' &= c_{11}\Delta x + c_{12}\Delta y + c_{13}\Delta z \,, \\
\Delta y' &= c_{21}\Delta x + c_{22}\Delta y + c_{23}\Delta z \,, \\
\Delta z' &= c_{31}\Delta x + c_{32}\Delta y + c_{33}\Delta z \,.
\end{aligned}\right\} \quad (3)$$

From equations (2) and (3) it follows that the coordinate differences transform so as to leave the expression for the distance between two points, Δr, fixed in value or *invariant*, that is, we have:

$$\Delta r^2 = \Delta x^2 + \Delta y^2 + \Delta z^2 = \Delta x'^2 + \Delta y'^2 + \Delta z'^2 \,. \quad (4)$$

A *vector* (in a three-dimensional continuum) may now be defined as a set of three quantities which transform from one rectangular Cartesian coordinate-system to another in the same manner as the coordinate differences, i.e., in accordance with equations analogous to equations (3). Thus (Δx, Δy, Δz) constitutes a vector. We shall designate vectors by sets of three letters representing their components (in some rectangular Cartesian coordinate-system) or, more briefly, by a single boldface letter. Invariant expressions, as in equation (4), are sometimes called *scalars*. The appropriateness of using vectors to represent physical magnitudes like velocity, acceleration, and force, which are surely independent of the particular set of spatial axes used in measuring them, is now evident.

Addition and *subtraction* of vectors may be defined in terms of the ordinary algebraic addition and subtraction of the respective components of the vectors; thus, if $A = (A_x, A_y, A_z)$ and $B = (B_x, B_y, B_z)$, then $A + B$ represents a new vector C with components ($C_x = A_x + B_x$, $C_y = A_y + B_y$, $C_z = A_z + B_z$). The *scalar* (or *dot*) *product* of A and B is a scalar defined as follows:

$$A \cdot B = A_x B_x + A_y B_y + A_z B_z.$$

The *vector* (or *cross*) *product* of A and B is a vector defined as follows:

$$A \times B = P,$$

where

$$P_x = A_y B_z - A_z B_y,$$

$$P_y = A_z B_x - A_x B_z,$$

$$P_z = A_x B_y - A_y B_x.$$

It is obvious that $A \times A = 0$. The scalar product of any vector with itself is, on the other hand, a scalar with the value:

$$A \cdot A = A^2 = A_x^2 + A_y^2 + A_z^2.$$

The sense in which vectors possess both "magnitude" and "direction" (which is often taken as the defining characteristic of vectors) should now be clear, namely, the *magnitude* of any vector A with components A_x, A_y, A_z is given by the absolute value of $(A \cdot A)^{1/2}$, i.e., $|A| = |(A_x^2 + A_y^2 + A_z^2)^{1/2}|$; and the *direction* of A is given by its three direction cosines (i.e., the cosines of the angles which A makes with the three coordinate axes), $A_x/|A|$, $A_y/|A|$, $A_z/|A|$. Like its square, $|A|$ is, of course, an invariant; the direction cosines are not invariants but from their values in a given coordinate-system their values may be calculated in any other specified coordinate-system.

I mention now two important vector operators, the "divergence" and the "curl" (or "rotation"). The *divergence* of any vector A is a scalar defined as:

$$\text{div } A = \frac{\partial A_x}{\partial x} + \frac{\partial A_y}{\partial y} + \frac{\partial A_z}{\partial z}, \tag{5}$$

where A_x, A_y, and A_z are respectively the x-, y-, and z-components of A.

The *curl* of any vector A is a vector whose x-, y-, and z-components respectively are defined as:

$$(\text{curl } A)_x = \left(\frac{\partial A_z}{\partial y} - \frac{\partial A_y}{\partial z} \right),$$

$$(\text{curl } A)_y = \left(\frac{\partial A_x}{\partial z} - \frac{\partial A_z}{\partial x} \right),$$

$$(\text{curl } A)_z = \left(\frac{\partial A_y}{\partial x} - \frac{\partial A_x}{\partial y} \right),$$

$$(6)$$

where A_x, A_y, and A_z are once again respectively the x-, y-, and z-components of A.

Tensors may be defined in a rather natural fashion as a generalization of the concept of vectors, namely, a (three-dimensional) *tensor* is a set of 3^N quantities which transform from one rectangular Cartesian coordinate-system to another in accordance with the appropriate analogue of equation (3). This defining condition can be stated more precisely with the help of an improved notation. From now on, the three coordinate axes will be designated x_1, x_2, x_3, so that all variable indices (subscripts or superscripts) which occur in a vector or tensor expression will be understood to take on all the values, 1, 2, 3. Furthermore, summation signs will be omitted in accordance with the convention that any variable indices which occur twice in a product are to be summed over all possible values. To illustrate, equations (3) will now be written:

$$\Delta x'_\mu = c_{\mu\nu} \Delta x_\nu \qquad (\mu = 1,\, 2,\, 3) . \quad (7)$$

Returning to our definition of a tensor, we see that a tensor is a set of 3^N quantities which transform as follows:

$$A'_{\mu\nu\sigma} \ldots = c_{\mu\alpha} c_{\nu\beta} c_{\sigma\gamma} \cdots A_{\alpha\beta\gamma} \ldots ,$$

where the c's are constant coefficients. The number of indices, N, is called the *rank* of the tensor. Vectors are obviously tensors of rank 1, and scalars may be considered tensors of rank 0. The so-called "Kronecker delta symbol," which occurs in equation (2), may be mentioned as a simple but highly important example of a tensor (of rank 2).

Instead of considering any further properties of tensors at this point, we proceed at once to the problem of defining tensors in the context of arbitrary coordinate transformations. We consider an n-dimensional continuous manifold in which a point is designated by an ordered class of n real numbers. A general curvilinear coordinate-system is any arbitrary assignment of such ordered classes of n real numbers to the points of the manifold such that one and only one class is assigned to each point. Two such coordinate-systems will always be connected by a coordinate transformation defined by n continuous differentiable functions with non-vanishing Jacobian (this last condition is to insure a one–one correspondence between the sets of coordinates assigned by the two coordinate-systems). If x_μ and x'_μ represent two such coordinate-systems, their respective coordinate differentials will be connected by the linear homogeneous equations:

$$d x'_\mu = \frac{\partial x'_\mu}{\partial x_\nu} d x_\nu \qquad (\mu = 1, 2, 3, \ldots n), \quad (8)$$

where the coefficients $\partial x'_\mu / \partial x_\nu$ are, in general, functions of the x_μ. On comparing equation (8) with equation (7), we see that the $\partial x'_\mu / \partial x_\nu$ are analogous to the $c_{\mu\nu}$. Hence we define a *contravariant n-vector*, by analogy with equation (7), as a set of n quantities A^μ ($\mu = 1, 2, 3, \ldots n$) which transform according to the law:

$$A'^\mu = \frac{\partial x'_\mu}{\partial x_\nu} A^\nu \qquad (\mu = 1, 2, 3, \ldots n). \quad (9)$$

Now, equation (9) suggests another slightly different law of transformation, involving once again only linear homogeneous equations, *viz.*:

$$A'_\mu = \frac{\partial x_\nu}{\partial x'_\mu} A_\nu \qquad (\mu = 1, 2, 3, \ldots n), \quad (10)$$

which we use to define a *covariant n-vector*. It should be noted that contravariant and covariant vectors have been distinguished by the use of superscript and subscript indices respectively—except in the case of the coordinates themselves, where (a) x_μ is not a vector at all, (b) dx_μ is a contravariant vector, and (c) $\partial / \partial x_\mu$ may be thought of as a covariant operator.

By generalizing the two preceding definitions we arrive at the concept of an *n*-tensor of rank N, which is a set of n^N quantities, $A^{\mu\nu\sigma\ldots}_{\alpha\beta\gamma\ldots}$, which transform with respect to each of the N indices like a vector (contravariant or covariant), i.e., by linear and homogeneous transformation equations. It is obvious that a tensor may be purely contravariant, purely covariant, or mixed. (Incidentally, in the case of rectangular Cartesian transformations the transformation coefficients $\partial x'_\nu / \partial x_\mu$ and $\partial x_\mu / \partial x'_\nu$ are identical [simply the c's of equation (2)], so that the distinction between contravariant and covariant tensors disappears.) A tensor of rank greater than 1 is *symmetrical* with respect to two indices μ and ν (both contravariant or both covariant) if the tensor remains unchanged when μ and ν are interchanged. A tensor of rank greater than 1 is *antisymmetrical* (or *skewsymmetrical*) with respect to a pair of indices μ and ν (both contravariant or both covariant) if the tensor changes sign when the μ and ν are interchanged. Both symmetry and antisymmetry are invariant properties of tensors.

Two tensors of equal rank and with equal numbers of indices of the same type may be added or subtracted to give a new tensor whose components consist of the ordinary algebraic sums or differences of corresponding components of the two original tensors. Two tensors of ranks M and N may be multiplied to give a new tensor of rank $M + N$. Also, a mixed tensor may be *contracted* by identifying two indices of different type and then summing over this index, the process issuing in a new tensor of rank 2 less than the original tensor.

A few examples of important tensors may be mentioned for purposes of illustration. Consider first a scalar field $\psi(x_1, x_2, \ldots x_n)$, where ψ is a function of the

x_μ which is invariant with respect to arbitrary coordinate transformations. For an infinitesimal displacement dx_μ, the corresponding change in ψ is given by:

$$d\psi = \frac{\partial \psi}{\partial x_\mu} d x_\mu. \tag{11}$$

The quantities $\partial\psi/\partial x_\mu$ (known as the *gradient* of ψ) transform into another coordinate-system as follows:

$$\frac{\partial \psi}{\partial x'_\nu} = \frac{\partial x_\mu}{\partial x'_\nu} \frac{\partial \psi}{\partial x_\mu} \qquad\qquad (\nu = 1, 2, 3 \ldots n),$$

which shows that the gradient of ψ is a covariant tensor of first order (or a covariant vector). Notice that in equation (11) the "product" of a contravariant vector, dx_μ, and a covariant vector, $\partial\psi/\partial x_\mu$, is an invariant, $d\psi$. It can be proved generally that the product of a contravariant and a covariant vector is an invariant (the proof follows almost immediately from the definitions [9] and [10]); thus we have found what corresponds to the scalar product of two vectors in orthogonal transformations.

Suppose we assume now that the manifold under consideration is Riemannian (or metrical), so that the infinitesimal "distance" (squared) between any two points is an invariant homogeneous quadratic function of the coordinate differentials:

$$d s^2 = g_{\mu\nu} d x_\mu d x_\nu, \tag{12}$$

where the $g_{\mu\nu}$ are, in general, functions of the x_μ. By a general theorem of tensor algebra, the product of any set of quantities $T^{\sigma\tau}_{\mu\nu}{:::}$ with the appropriate number of arbitrarily selected contravariant and covariant vectors,

$$T^{\sigma\tau}_{\mu\nu}{\cdots} A_\sigma B_\tau \ldots F^\mu G^\nu \ldots,$$

is an invariant if, and only if, $T^{\sigma\tau}_{\mu\nu}{:::}$ is a tensor. Hence, it follows immediately from equation (12) that $g_{\mu\nu}$ constitutes a covariant, symmetrical tensor of rank 2 (the metric tensor). Also, a contravariant tensor $g^{\mu\nu}$ may be defined in terms of $g_{\mu\nu}$, as follows:

$$g^{\mu\nu} = \frac{\text{minor} \mid g_{\mu\nu} \mid}{g},$$

where minor $\mid g_{\mu\nu} \mid$ is the minor of the determinant formed by the $g_{\mu\nu}$ and $g = \mid g_{\mu\nu} \mid$ is the full determinant formed by the $g_{\mu\nu}$.

By means of the general theorem of tensor algebra stated above, one can easily show that the Kronecker delta symbol designates a mixed tensor of rank 2, written δ^μ_ν. Also, this tensor may be used to express the relationship between the transformation coefficients of contravariant and covariant vectors, as follows:

$$\frac{\partial x'_a}{\partial x_\nu} \frac{\partial x_\mu}{\partial x'_a} = \delta^\mu_\nu. \tag{13}$$

Equation (2) is just a special case of equation (13). Since

$$g_{\mu a} g^{\nu a} = \delta_{\mu}^{\nu} ,$$

the tensor δ_{μ}^{ν} is sometimes also written g_{μ}^{ν}. In terms of the tensors $g_{\mu\nu}$, $g^{\mu\nu}$, and g_{μ}^{ν}, one may define general processes for raising, lowering, and changing indices in a given tensor:

$$A^{\nu} = g^{\nu a} A_a ,$$

$$A_{\mu} = g_{\mu a} A^a ,$$

$$A^{\nu} = g_a^{\nu} A^a .$$

Tensors may be differentiated to give new tensors but the process of differentiation must be a generalization of ordinary differentiation; in other words, the ordinary differentiation of a tensor does not in general produce a tensor, so that we must define a new type of differentiation (called "covariant differentiation") which always produces tensors from given tensors. The appropriate definition clearly must take account of the changing values of the transformation coefficients of a given tensor as one moves from point to point in a Riemannian manifold; and one would expect these changes to be a function of the metrical properties of the manifold, i.e., the values of the $g_{\mu\nu}$. It turns out that *covariant differentiation* (symbolized by a subscript index preceded by a semicolon) may be defined as follows, for contravariant and covariant tensors, respectively:

$$A^{\mu}_{;\nu} = \frac{\partial A^{\mu}}{\partial x_{\nu}} + \{ a\nu, \mu \} A^a , \tag{14}$$

$$A_{\mu;\nu} = \frac{\partial A_{\mu}}{\partial x_{\nu}} - \{ \mu\nu, a \} A_a , \tag{15}$$

where

$$\{ \mu\nu, \tau \} = \tfrac{1}{2} g^{\tau a} \left(\frac{\partial g_{\mu a}}{\partial x_{\nu}} + \frac{\partial g_{\nu a}}{\partial x_{\mu}} - \frac{\partial g_{\mu\nu}}{\partial x_a} \right).$$

Equations (14) and (15) may be generalized for tensors of ranks higher than 1. It should be noted that covariant differentiation always increases the rank of a tensor by one. The sense in which covariant differentiation is a generalization of ordinary differentiation is easily seen: when $\{ \mu\nu, \tau \}$ vanishes, the two types of differentiation become identical (this occurs when all the components of the metrical tensor $g_{\mu\nu}$ are constants, i.e., in the case of rectangular Cartesian coordinates). If rectangular Cartesian coordinates can be introduced into a given Riemannian manifold, the latter is said to be Euclidean (or pseudo-Euclidean). The condition that such a rectangular Cartesian coordinate-system exists is equivalent to the vanishing of a certain fourth rank tensor, the Riemann-Christoffel tensor:

$$R_{\mu\nu\sigma}^{\rho} = \{ \mu\sigma, a \} \{ a\nu, \rho \} - \{ \mu\nu, a \} \{ a\sigma, \rho \} + \frac{\partial}{\partial x_{\nu}} \{ \mu\sigma, \rho \} - \frac{\partial}{\partial x_{\sigma}} \{ \mu\nu, \rho \} .$$

The large subject of tensor analysis cannot be entered into here but in order to illustrate the power of tensor analytical methods we may consider the method devised by Minkowski for expressing the Maxwell-Lorentz equations in generally covariant form. Let ϕ_ν be a covariant 4-vector, the "electromagnetic potential vector." From ϕ_ν one can construct an antisymmetrical 4-tensor of second rank, $F_{\rho\sigma}$, as follows:

$$F_{\rho\sigma} = \frac{\partial \phi_\rho}{\partial x_\sigma} - \frac{\partial \phi_\sigma}{\partial x_\rho}. \tag{16}$$

Now, an antisymmetrical tensor of second rank has just six independent components (since $F_{\rho\sigma} = 0$ for $\rho = \sigma$, and $F_{\rho\sigma} = -F_{\sigma\rho}$ for $\rho \neq \sigma$), which can be identified, in sets of three, with respectively the electric field intensity and the magnetic field intensity. The explicit identification is made by noting that $F_{\rho\sigma}$ satisfies the equation:

$$\frac{\partial F_{\rho\sigma}}{\partial x_\tau} + \frac{\partial F_{\sigma\tau}}{\partial x_\rho} + \frac{\partial F_{\tau\rho}}{\partial x_\sigma} = 0. \tag{17}$$

(Terms such as those in [14] and [15] containing the metric tensor do not appear in equations [16] and [17] because they cancel out.) The left-hand side of (17) is an antisymmetrical tensor of third rank, consisting therefore of just four independent components (one component for each triple of values [ρ, σ, τ], such that ρ, σ, $\tau = 1, 2, 3, 4$, and $\rho \neq \sigma$, $\rho \neq \tau$, $\sigma \neq \tau$). Equation (17), then, is equivalent to four distinct equations, *viz.*:

$$\left.\begin{aligned}
\frac{\partial F_{23}}{\partial x_4} + \frac{\partial F_{34}}{\partial x_2} + \frac{\partial F_{42}}{\partial x_3} &= 0, \\[2mm]
\frac{\partial F_{34}}{\partial x_1} + \frac{\partial F_{41}}{\partial x_3} + \frac{\partial F_{13}}{\partial x_4} &= 0, \\[2mm]
\frac{\partial F_{41}}{\partial x_2} + \frac{\partial F_{12}}{\partial x_4} + \frac{\partial F_{24}}{\partial x_1} &= 0, \\[2mm]
\frac{\partial F_{12}}{\partial x_3} + \frac{\partial F_{23}}{\partial x_1} + \frac{\partial F_{31}}{\partial x_2} &= 0.
\end{aligned}\right\} \tag{18}$$

Introducing the definitions

$$\begin{aligned}
F_{23} &= H_x, & F_{14} &= E_x, \\
F_{31} &= A_y, & F_{24} &= E_y, \\
F_{12} &= H_z, & F_{34} &= E_z,
\end{aligned}$$

we obtain from equations (18), using the definitions of curl and div,

$$\left.\begin{aligned}
\operatorname{curl} E + \frac{\partial H}{\partial t} &= 0, \\[2mm]
\operatorname{div} H &= 0
\end{aligned}\right\} \tag{19}$$

these are essentially identical with two of the Maxwell-Lorentz equations, the second and third respectively of equations (8.27) (the absence of the factor $1/c$ from the first of equations [19] is due to the choice of space and time units so that $c = 1$).

To obtain the remaining two Maxwell-Lorentz equations we introduce the contravariant tensor associated with $F_{\alpha\beta}$:

$$F^{\mu\nu} = g^{\mu\alpha} g^{\nu\beta} F_{\alpha\beta} , \tag{20}$$

and the contravariant vector I^μ of the electric current density. It can then be shown that the following equation holds:

$$\frac{\partial F^{\mu\nu}}{\partial x_\nu} = I^\mu . \tag{21}$$

Introducing the definitions:

$$F^{23} = H'_x , \qquad F^{14} = -E'_x ,$$

$$F^{31} = H'_y , \qquad F^{24} = -E'_y ,$$

$$F^{12} = H'_z , \qquad F^{34} = -E'_z ,$$

$$I^1 = 4\pi i_x, \qquad I^2 = 4\pi i_y , \qquad I^3 = 4\pi i_z , \qquad I^4 = 4\pi \rho ,$$

we obtain from equation (21), using the definitions of curl and div,

$$\left. \begin{array}{l} \text{div } E' = 4\pi \rho , \\[2mm] \text{curl } H' - \dfrac{\partial E}{\partial t} = 4\pi i . \end{array} \right\} \tag{22}$$

These are essentially identical with the remaining two Maxwell-Lorentz equations, the first and fourth respectively of equations (8.27) (once more, the absence of the factor $1/c$ is due to the choice of space and time units). The three sets of equations (17), (20), and (21) represent a generalization of the Maxwell-Lorentz field equations, since E is identical with E' and H with H', in general, only for the coordinate-systems of special relativity.

For an introduction to tensor analysis reference may be made to P. Bergmann, *Introduction to the Theory of Relativity*, chap. v, and to J. L. Synge and A. Schild, *Tensor Calculus*.

SURVEY OF

WRITINGS ON

WHITEHEAD'S

THEORY OF

RELATIVITY

The following list includes, as far as I have been able to ascertain, all writings since the publication of R in 1922 concerned primarily with the physical (rather than the philosophical) aspects of Whitehead's theory of relativity. (Full references will be found in the Bibliography.)

[1] G. Temple, "A Generalisation of Professor Whitehead's Theory of Relativity" (1923).

[2] A. S. Eddington, "A Comparison of Whitehead's and Einstein's Formulae" (1924).

[3] G. Temple, "Central Orbits in Relativistic Dynamics Treated by the Hamilton-Jacobi Method" (1924).

[4] W. Band, "Dr. A. N. Whitehead's Theory of Absolute Acceleration" (1929).

[5] W. Band, "A Comparison of Whitehead's with Einstein's Law of Gravitation" (1929).

[6] W. Band, "Is Space-Time Flat?" (1942).

[7] J. L. Synge, *The Relativity Theory of A. N. Whitehead* (1951).

[8] J. L. Synge, "Orbits and Rays in the Gravitational Field of a Finite Sphere According to the Theory of A. N. Whitehead" (1952).

[9] G. L. Clark, "The Problem of Two Bodies in Whitehead's Theory" (1953).

10] C. B. Rayner, "The Application of the Whitehead Theory of Relativity to Non-static, Spherically Symmetrical Systems" (1954).

.11] J. L. Synge, "Note on the Whitehead-Rayner Expanding Universe" (1954).

[12] C. B. Rayner, "The Effects of Rotation of the Central Body on its Planetary Orbits, after the Whitehead Theory of Gravitation" (1955).

[13] C. B. Rayner, "Whitehead's Law of Gravitation in a Space-Time of Constant Curvature" (1955).

[14] A. Schild, "On Gravitational Theories of Whitehead's Type" (1956).

There is also a thesis by C. B. Rayner, "Foundations and Applications of Whitehead's Theory of Relativity" (University of London, 1953), which I have not seen.

In [1] Temple generalizes Whitehead's theory of relativity on the basis of space-time with uniform and isotropic curvature (but see [12]). The generalized theory is shown to lead to observational predictions for the advance of perihelion of Mercury and for the deviation of light-rays passing near the sun, which are not detectably different from the predictions of Whitehead's original theory.

In [2] Eddington proves that both Einstein's and Whitehead's theories of relativity give identical expressions (the Schwarzschild solution) for the field of a mass-point at rest.

In [3] Temple first shows how to treat problems in relativistic dynamics by the Hamilton-Jacobi method of classical dynamics. He then applies this method to the problems of planetary motion and the motion of light-rays in a gravitational field according to Einstein's and Whitehead's theories of relativity. Einstein's and Whitehead's respective laws of gravitation are shown to lead to identical solutions for these two problems. (This, of course, simply confirms Eddington's result in [2].) Furthermore, Temple studies one of the two laws of gravitation that Whitehead proposes as alternatives to both his own and Einstein's laws (see R, p. 86, equation [21]). On the basis of this new law of gravitation the solutions to the planetary and light-ray problems mentioned above are shown to involve an arbitrary function, which may be so adjusted as to yield exactly the same results as Einstein's and Whitehead's laws.

In [4] Band criticizes as illogical Whitehead's acceptance of the dependence of space-time coordinate-systems on velocity but not on acceleration. Band also claims to derive from the purely kinematic side of Whitehead's theory an empirical consequence which is obviously false (but see [6]).

In [5] Band discusses and extends the results in [3] using a somewhat different approach from Temple's. Band shows that Whitehead's own law of gravitation does not give empirically correct predictions for orbits of high eccentricity (e.g., motions of comets), *if* one rejects Whitehead's view and interprets the metric as dependent on the gravitational field. As for Whitehead's alternative law of gravitation, Band concludes that because it presupposes the independence of metric and gravitational field, it is inferior to Einstein's law which requires no such presupposition.

In [6] Band states some general criticisms of the principle (not limited to Whitehead's theory of relativity) that space-time is flat. He argues that this principle entails the assumption of a standard of absolutely uniform motion, and that in practice the use of flat space-time, while not incorrect, is extremely inconvenient.

In [7] Synge provides first of all a compact mathematical formulation of Whitehead's laws of motion, gravitation, and electromagnetism. Synge's notation and terminology are closer than Whitehead's to those in current use, and therefore Synge's analysis may be easier to follow than Whitehead's. Synge gives derivations of the main empirical consequences of Whitehead's theory which are more detailed—and to that extent clearer—than those of Whitehead himself. Synge also extends Whitehead's law of gravitation to the case of continuous static distributions of matter, and then calculates the gravitational field of a finite sphere of uniform density at rest.

In [8] Synge calculates the gravitational field of a finite sphere with spherically symmetrical density at rest, and then investigates orbits and light-rays in such a field. The latter results when applied to the solar field are shown to be in good agreement with the tests of general relativity on the perihelion of Mercury and on the deflections of light-rays passing near the sun.

In [9] Clark investigates the motion of the center of mass of a two-body system according to Whitehead's laws of motion and gravitation. Clark proves that there is a secular acceleration of the center of mass, a result which disagrees with the corresponding result in general relativity. Accordingly, Clark suggests that a choice between Einstein's and Whitehead's theories might be made on empirical grounds by observing the motions of the centers of mass of double stars.

In [10] Rayner first extends Whitehead's law of gravitation to non-static continuous distributions of matter and then on the basis of this extension proceeds to construct a uniformly expanding, homogeneous, isotropic cosmological model. This same model is derived in a simpler way by Synge in [11].

In [12] Rayner generalizes Synge's results in [8] by calculating the gravitational field of a finite, uniformly rotating, homogeneous sphere and then determining the perturbing forces on the Newtonian elliptical orbits due to rotation of the central sphere. The results are shown to agree with those obtained by Lense and Thirring (1918) using the linear approximational form of Einstein's law of gravitation.

In [13] Rayner first indicates an error in Temple's results in [1] and then goes on to solve the problem of formulating Whitehead's law of gravitation in a space-time of constant curvature by a method different from Temple's. This result is then extended to include systems with continuous proper-density and velocity distributions. Rayner suggests that perhaps the most appropriate application of the generalized law of gravitation which he derives is to a cosmological theory like that in [10].

In [14] Schild shows that there is an infinity of gravitational theories, which are generalizations of Whitehead's theory, employing only *flat* space-time and agreeing with all observational data presently available. Schild suggests that all these theories probably exhibit the secular acceleration of the center of mass of a two-body system found by Clark in Whitehead's original theory. Hence observations on double stars may permit an empirical decision between general relativity and all gravitational theories of Whitehead's type.

BIBLIOGRAPHY

AGAR, W. E. *A Contribution to the Theory of the Living Organism.* 2d ed. rev. Carlton, Australia: Melbourne University Press, 1951.

BAND, WILLIAM. "A Comparison of Whitehead's with Einstein's Law of Gravitation," *Philosophical Magazine,* VII, No. 47 (1929), 1183–86.

———. "Dr. A. N. Whitehead's Theory of Absolute Acceleration," *ibid.,* No. 43 (1929), pp. 434–40.

———. "Is Space-Time Flat?" *Physical Review,* LXI (1942), 698–701.

BERGMANN, PETER G. *Introduction to the Theory of Relativity.* New York: Prentice-Hall, Inc., 1942.

BIRKHOFF, G. D. "Matter, Electricity, and Gravitation in Flat Space-Time," *Proceedings of the National Academy of Sciences,* XXIX (1943), 231–39.

BONOLA, ROBERTO. *Non-Euclidean Geometry.* Translated by H. S. CARSLAW. Chicago: Open Court, 1912. (Reprinted, New York: Dover Publications, 1955.)

BROAD, C. D. "Critical Notices: *The Principles of Natural Knowledge* by A. N. Whitehead," *Mind,* XXIX (1920), 216–31.

———. "Critical Notices: *The Principle of Relativity* by A. N. Whitehead," *ibid.,* XXXII (1923), 211–19.

CLARK, G. L. "The Problem of Two Bodies in Whitehead's Theory," *Proceedings of the Royal Society of Edinburgh,* Ser. A, LXIV (1954), 49–56.

COXETER, H. S. M. *Non-Euclidean Geometry.* 3d ed. Toronto: University of Toronto Press, 1957.

DE BROGLIE, LOUIS. *The Revolution in Physics.* Translated by RALPH NIEMEYER. New York: Noonday Press, 1953.

DE LAGUNA, THEODORE. "Point, Line, and Surface, as Sets of Solids," *Journal of Philosophy,* XIX (1922), 449–61.

EDDINGTON, A. S. "A Comparison of Whitehead's and Einstein's Formulae," *Nature* (London), CXIII (1924), 192.

———. *The Mathematical Theory of Relativity.* 2d ed. Cambridge: Cambridge University Press, 1924.

———. *The Nature of the Physical World.* New York: Macmillan Co., 1928.

———. *Space, Time, and Gravitation.* Cambridge: Cambridge University Press, 1920.

EINSTEIN, ALBERT. *The Meaning of Relativity.* 5th ed. rev. Princeton: Princeton University Press, 1955.

———. "Physik und Realität," *Journal of the Franklin Institute,* CCXXI (1936), 313–47.

EINSTEIN, ALBERT. *Relativity, the Special and the General Theory*. Translated by ROBERT W. LAWSON. New York: Peter Smith, 1931.

———. *Sidelights on Relativity*. Translated by G. B. JEFFERY and W. PERRETT. New York: E. P. Dutton & Co., 1922.

EINSTEIN, ALBERT, and INFELD, LEOPOLD. "The Gravitational Equations and the Problem of Motion, II," *Annals of Mathematics*, XLI (1940), 455–64.

EINSTEIN, ALBERT; INFELD, LEOPOLD; and HOFFMANN, BANESH. "The Gravitational Equations and the Problem of Motion," *Annals of Mathematics*, XXXIX (1938), 65–100.

FORDER, H. G. *Geometry*. London: Hutchinson's University Library, 1950.

FRANK, PHILIPP. *Einstein, His Life and Times*. Translated by GEORGE ROSEN; edited by SHUICHI KUSAKA. New York: Alfred A. Knopf, 1947.

FREUNDLICH, ERWIN. *The Foundations of Einstein's Theory of Gravitation*. Translated by HENRY L. BROSE. New York: G. E. Stechert & Co., 1922.

GÖDEL, KURT. "An Example of a New Type of Cosmological Solutions of Einstein's Field Equations of Gravitation," *Reviews of Modern Physics*, XXI (1949), 447–50.

———. "A Remark about the Relationship between Relativity Theory and Idealistic Philosophy," *Albert Einstein: Philosopher-Scientist*. Edited by P. A. SCHILPP. Evanston, Ill.: Library of Living Philosophers, 1949. (Now published: New York: Tudor Publishing Co.)

GOLDSTEIN, HERBERT. *Classical Mechanics*. Cambridge, Mass.: Addison-Wesley Press, 1950.

GRÜNBAUM, ADOLF. "Conventionalism in Geometry," *The Axiomatic Method with Special Reference to Geometry and Physics*. Edited by L. HENKIN, P. SUPPES, and A. TARSKI. Amsterdam: North-Holland Publishing Co., 1959.

———. "The Philosophical Retention of Absolute Space in Einstein's General Theory of Relativity," *Philosophical Review*, LXVI (1957), 525–34.

HUNTINGTON, EDWARD V. "The Duplicity of Logic," *Scripta Mathematica*, V (1938), 149–57, 233–38.

———. "A Set of Postulates for Abstract Geometry, Expressed in Terms of the Simple Relation of Inclusion," *Math. Annalen*, LXXIII (1913), 522–59.

JAMES, WILLIAM. *Principles of Psychology*, Vol. I. New York: Henry Holt & Co., 1890.

JOOS, GEORG. *Theoretical Physics*. 3d ed. London: Blackie & Son, 1958.

KLEIN, FELIX. *Elementary Mathematics from an Advanced Standpoint*, Vol. II: *Geometry*. Translated by E. R. HEDRICK and C. A. NOBLE. New York: Macmillan Co., 1939. (Reprinted, New York: Dover Publications, n.d.)

KRAICHMAN, ROBERT H. "Special-Relativistic Derivation of Generally Covariant Gravitation Theory," *Physical Review*, XCVIII (1955), 1118–22.

LEWIS, C. I. "Whitehead and the Categories of Natural Knowledge," *The Philosophy of Alfred North Whitehead*. 2d ed. rev. Edited by P. A. SCHILPP. New York: Tudor Publishing Co., 1951.

LILLIE, RALPH STAYNER. *General Biology and Philosophy of Organism*. Chicago: University of Chicago Press, 1945.

LORENTZ, H. A., *et al*. *The Principle of Relativity*. Translated by W. PERRETT and G. B. JEFFERY. London: Methuen & Co., 1923. (Reprinted, New York: Dover Publications, n.d.)

MACAULAY, W. H. "Newton's Theory of Kinetics," *Bulletin of the American Mathematical Society*, III (1897), 363–71.

McCREA, W. H. "On the Objective of Einstein's Work," *British Journal for the Philosophy of Science*, VIII (1957–58), 18–29.

MENGER, K. "Topology without Points," *Rice Institute Pamphlet*, XXVII (1940), 80–107.

MØLLER, C. *The Theory of Relativity*. Oxford: Clarendon Press, 1952.

NORTHROP, F. S. C. "Einstein's Conception of Science," *Albert Einstein: Philosopher-Scientist*. Edited by P. A. SCHILPP. Evanston, Ill.: Library of Living Philosophers, 1949. (Now published: New York: Tudor Publishing Co.)

PALTER, ROBERT. "Philosophic Principles and Scientific Theory," *Philosophy of Science*, XXIII (1956), 111–35.

PAULI, WOLFGANG. *Theory of Relativity*. Translated by G. FIELD. New York: Pergamon Press, 1958.

POINCARÉ, HENRI. *Science and Hypothesis* in *The Foundations of Science*. Translated by G. B. HALSTED. Lancaster, Pa.: Science Press, 1946.

RAYNER, C. B. "The Application of the Whitehead Theory of Relativity to Nonstatic, Spherically Symmetrical Systems," *Proceedings of the Royal Society of London*, Ser. A, CCXXII (1954), 509–26.

———. "The Effects of Rotation of the Central Body on Its Planetary Orbits, after the Whitehead Theory of Gravitation," *ibid.*, CCXXXII (1955), 135–48.

———. "Whitehead's Law of Gravitation in a Space-Time of Constant Curvature," *Proceedings of the Physical Society of London*, Section B, LXVIII, Part II (1955), 944–50.

REICHENBACH, HANS. *Axiomatik der relativistischen Raum-Zeit-Lehre*. Brunswick: Druck und Friedr. Vieweg & Sohn, 1924.

———. *Philosophie der Raum-Zeit-Lehre*. Berlin: Walter de Gruyter & Co., 1928. (Translated by MARIA REICHENBACH and JOHN FREUND as *The Philosophy of Space and Time*. New York: Dover Publications, 1957.)

RIEMANN, BERNHARD. "Über die Hypothesen, Welche der Geometrie zu Grunde liegen" (1854), *Gesammelte Mathematische Werke* (pp. 272–87). Edited by H. WEBER and R. DEDEKIND. Leipzig: Druck und Verlag von B. G. Teubner, 1892; reprinted, New York: Dover Publications, 1953. (Translated as "On the Hypotheses Which Lie at the Bases of Geometry," in WILLIAM KINGDON CLIFFORD, *Mathematical Papers* [pp. 55–71]. Edited by ROBERT TUCKER. London: Macmillan & Co., 1882.)

ROBINSON, GILBERT DE B. *The Foundations of Geometry*. 2d ed. rev. Toronto: University of Toronto Press, 1946.

RUSSELL, BERTRAND A. W. *An Essay on the Foundations of Geometry*. Cambridge: Cambridge University Press, 1897. (Reprinted, New York: Dover Publications, 1956.)

SCHILD, A. "On Gravitational Theories of Whitehead's Type," *Proceedings of the Royal Society of London*, Ser. A, CCXXXV (1956), 202–9.

SCHRÖDINGER, ERWIN. *Space-Time Structure*. Cambridge: Cambridge University Press, 1950.

SCOTT, CHARLOTTE ANGAS. "On Cayley's Theory of the Absolute," *Bulletin of the American Mathematical Society*, III (1897), 235–46.

SILBERSTEIN, LUDWIG. *The Theory of Relativity*. London: Macmillan & Co., 1924.

STRUIK, DIRK J. *Lectures on Classical Differential Geometry*. Reading, Mass.: Addison-Wesley Publishing Co., 1950.

SYNGE, J. L. "Note on the Whitehead-Rayner Expanding Universe," *Proceedings of the Royal Society of London*, Ser. A, CCXXVI (1954), 336–38.

———. "Orbits and Rays in the Gravitational Field of a Finite Sphere According to the Theory of A. N. Whitehead," *ibid.*, CCXI (1952), 303–19.

———. *The Relativity Theory of A. N. Whitehead.* Lecture Series 5, Institute for Fluid Dynamics and Applied Mathematics, University of Maryland, 1951.

———. *Relativity: The Special Theory.* New York: Interscience Publishers, 1956.

SYNGE, J. L., and SCHILD, A. *Tensor Calculus.* Toronto: University of Toronto Press, 1949.

TEMPLE, G. "Central Orbits in Relativistic Dynamics Treated by the Hamilton-Jacobi Method," *Philosophical Magazine*, 6th ser., XLVIII, No. 284 (1924), 277–92.

———. "A Generalisation of Professor Whitehead's Theory of Relativity," *Proceedings of the Physical Society of London*, XXXVI (1923–24), 176–93.

TOLMAN, RICHARD C. *Relativity, Thermodynamics, and Cosmology.* Oxford: Clarendon Press, 1934.

VEBLEN, OSWALD. "The Foundations of Geometry," *Monographs on Topics of Modern Mathematics.* Edited by J. W. A. YOUNG. New York: Longmans, Green & Co., 1911. (Reprinted, New York: Dover Publications, 1955.)

WEYL, HERMANN. *Raum-Zeit-Materie.* Berlin: J. Springer, 1921. (Translated by HENRY L. BROSE as *Space-Time-Matter* [1922]. Reprinted, New York: Dover Publications, 1950.)

WHITEHEAD, ALFRED NORTH. *Adventures of Ideas.* New York: Macmillan Co., 1933.

———. *The Aims of Education and Other Essays.* London: Williams & Norgate, 1932.

———. "Analysis of Meaning" from "Remarks," *Philosophical Review*, XLVI (1937), 178–86. (Reprinted in *Essays in Science and Philosophy.*)

———. "The Anatomy of Some Scientific Ideas," *The Organisation of Thought, Educational and Scientific.* London: Williams & Norgate, 1917. (Reprinted with some omissions as chap. ix of *The Aims of Education.*)

———. *The Axioms of Descriptive Geometry.* Cambridge: Cambridge University Press, 1907.

———. "The Axioms of Geometry," from "Geometry," *Encyclopaedia Britannica*, 11th ed. (1910), XI, 730–36. (Reprinted without diagrams in *Essays in Science and Philosophy.*)

———. *The Axioms of Projective Geometry.* Cambridge: Cambridge University Press, 1906.

———. *The Concept of Nature.* Cambridge: Cambridge University Press, 1920.

———. "Einstein's Theory," *The* (London) *Times Educational Supplement*, February 12, 1920, p. 83. (Reprinted in *Essays in Science and Philosophy.*)

———. *An Enquiry Concerning the Principles of Natural Knowledge.* 2d ed. rev. Cambridge: Cambridge University Press, 1925.

———. *Essays in Science and Philosophy.* New York: Philosophical Library, 1948.

———. "The Idealistic Interpretation of Einstein's Theory," *Proceedings of the Aristotelian Society*, N.S., XXII (1921–22), 130–34. (Whitehead's contribution to a discussion to which H. Wildon Carr, T. P. Nunn, and Dorothy Wrinch also contributed.)

———. "Letter to the Editor," *New Statesman*, XX (February 17, 1923), 568. (Reply to a review of *The Principle of Relativity.*)

———. *Modes of Thought.* New York: Macmillan Co., 1938.

————— (with BERTRAND RUSSELL). "Non-Euclidean Geometry," from "Geometry," *Encyclopaedia Britannica*, 11th ed. (1910), XI, 724–30. (Reprinted without diagrams in *Essays in Science and Philosophy*.)

—————. "On Mathematical Concepts of the Material World," *Philosophical Transactions, Royal Society of London*, Ser. A, CCV (1906), 465–525.

—————. "The Philosophical Aspects of the Theory of Relativity," *Proceedings of the Aristotelian Society*, N.S., XXII (1921–22), 215–23.

—————. *The Principle of Relativity*. Cambridge: Cambridge University Press, 1922.

—————. "The Problem of Simultaneity," *Relativity, Logic, and Mysticism*, pp. 34–41. Aristotelian Society Supplementary Vol. III. London: Williams & Norgate, 1923. (Whitehead's contribution to a symposium to which H. Wildon Carr and R. A. Sampson also contributed.)

—————. *Process and Reality*. New York: Macmillan Co., 1929.

—————. *Science and the Modern World*. New York: Macmillan Co., 1925.

—————. *Symbolism*. Cambridge: Cambridge University Press, 1928.

—————. "Time, Space, and Material: Are They, and If So in What Sense, the Ultimate Data of Science?" *Problems of Science and Philosophy*, pp. 44–57. Aristotelian Society Supplementary Vol. II. London: Williams & Norgate, 1919. (Whitehead's contribution to a symposium to which Sir Oliver Lodge, J. W. Nicholson, Henry Head, Mrs. Adrian Stephen, and H. Wildon Carr also contributed.)

—————. *A Treatise on Universal Algebra*. Cambridge: Cambridge University Press, 1898.

—————. "Uniformity and Contingency," *Proceedings of the Aristotelian Society*, N.S., XXIII (1922–23), 1–18. (Reprinted in *Essays in Science and Philosophy*.)

WHITTAKER, E. T. *From Euclid to Eddington*. Cambridge: Cambridge University Press, 1949.

—————. *A History of the Theories of Aether and Electricity*, Vol. II. London: Thomas Nelson & Sons, 1953.

—————. *A Treatise on the Analytical Dynamics of Particles and Rigid Bodies*. 4th ed. rev. Cambridge: Cambridge University Press, 1937. (Reprinted, New York: Dover Publications, 1944.)

WOODS, FREDERICK S. "Forms of Non-Euclidean Space," *Lectures on Mathematics*, pp. 31–74. The American Mathematical Society, Boston Colloquium. New York: Macmillan Co., 1905.

INDEX